Lecture Notes in Computer Science 16240

The series Lecture Notes in Computer Science (LNCS), including its subseries Lecture Notes in Artificial Intelligence (LNAI) and Lecture Notes in Bioinformatics (LNBI), has established itself as a medium for the publication of new developments in computer science and information technology research, teaching, and education.

LNCS enjoys close cooperation with the computer science R & D community, the series counts many renowned academics among its volume editors and paper authors, and collaborates with prestigious societies. Its mission is to serve this international community by providing an invaluable service, mainly focused on the publication of conference and workshop proceedings and postproceedings. LNCS commenced publication in 1973.

Abdelkader Hameurlain · A Min Tjoa ·
Mirian Halfeld Ferrari · Dario Colazzo

Editors

Transactions on Large-Scale Data- and Knowledge- Centered Systems LIX

Special Issue on Data Management - Principles, Technologies, and Applications

 Springer

Editors-in-Chief

Abdelkader Hameurlain
Toulouse University
Toulouse, France

A Min Tjoa
Vienna University of Technology
Vienna, Austria

Editor

Mirian Halfeld Ferrari
LIFO, University of Orleans
Orleans, France

Dario Colazzo
LAMSADE, University Paris Dauphine - PSL
Paris, France

ISSN 0302-9743   ISSN 1611-3349 (electronic)
Lecture Notes in Computer Science
ISSN 1869-1994   ISSN 2510-4942 (electronic)
Transactions on Large-Scale Data- and Knowledge-Centered Systems
ISBN 978-3-662-72448-4   ISBN 978-3-662-72449-1 (eBook)
https://doi.org/10.1007/978-3-662-72449-1

This Springer imprint is published by the registered company Springer-Verlag GmbH, DE,
part of Springer Nature.
The registered company address is: Heidelberger Platz 3, 14197 Berlin, Germany

If disposing of this product, please recycle the paper.

Preface

This volume includes a selection of thoroughly revised papers from the 40th Conference on Data Management: Principles, Technologies and Applications (BDA 2024). The special issue presents four articles on current research topics in data management, including graph optimization for information exposure, sequence similarity search, multi-model systems, and data privacy. Authors were invited to prepare extended journal versions of their contributions, which were fully re-reviewed by the editorial board of this issue.

We warmly thank all authors and the editorial board for their efforts and valuable contributions in enhancing the final versions of the papers.

Finally, we are grateful to the Editors-in-Chief, Abdelkader Hameurlain and A Min Tjoa, for giving us the opportunity to publish special issue as part of the TLDKS journal series.

September 2025
Mirian Halfeld Ferrari
Dario Colazzo

Organization

Editors-in-Chief

Abdelkader Hameurlain   Toulouse University, France
A Min Tjoa   TU Wien, Austria

Guest Editors

Mirian Halfeld Ferrari   LIFO, University of Orleans, France
Dario Colazzo   LAMSADE, University Paris Dauphine - PSL, France

Editorial Board

Contents

Maximizing Diverse Information Exposure in Content-Based Social
Networks 1
    Jonathan Colin and Silviu Maniu

SISIS: Sequence Indexing for SImilarity Search 32
    Sara Jarrad, Hubert Naacke, and Stéphane Gançarski

A Categorical Representation of Multi-model Data to Prevent Data
Migration Mismatch 61
    Annabelle Gillet and Éric Leclercq

Consistently Mapping Differential Privacy Paradigms Between Relational
Databases and RDF 94
    Sara Taki, Adrien Boiret, Cédric Eichler, and Benjamin Nguyen

Author Index 123

Maximizing Diverse Information Exposure in Content-Based Social Networks

Jonathan Colin[1]([✉]) and Silviu Maniu[2]

[1] Université Paris-Saclay, CNRS, LISN, Gif-sur-Yvette, France
`jonathan.colin@universite-paris-saclay.fr`
[2] Université Grenoble Alpes, CNRS, Grenoble INP, LIG, Grenoble, France
`silviu.maniu@univ-grenoble-alpes.fr`

Abstract. Online social networks have transformed communication, serving as key platforms for sharing and consuming information. These networks expose users to a range of opinions, either intentionally or incidentally. However, recommendation systems often favor similar content, overshadowing diverse, niche, or novel perspectives. This bias exacerbates challenges such as fake news, filter bubbles, and opinion polarization.

This paper introduces a framework to promote diversity in content-based social networks by framing *information exposure diversity* as an optimization problem. We focus on locally modifying the user-content graph by adding edges to maximize a diversity metric from an individual users' perspective. Importantly, we define diversity for two semantics: user-user and user-item recommendations. We formalize the concept of information exposure, linking it to established models in the literature, and propose several algorithms to address this problem, including gradient descent-based and greedy methods. Experiments on various real world datasets show that our algorithms are better than state-of-the-art methods in achieving higher diversity.

Keywords: graph optimization · diversity · polarization · social networks · graph data mining

1 Introduction

Social media is now ubiquitous in our society and has become the main source of information for an increasing number of people; in this sense, it holds a major role in the broadcast of information [28]. At the same time, exposure to social media effectively influences our opinions either *actively* (when we consume media) or even *passively* (when we browse content). This means that any content that a user merely *sees* (or is *exposed* to) also influences their beliefs, giving undue power to the algorithms that recommend content on online platforms, whose aim is to *maximize user engagement* – by only providing content that is similar to the users' current interests, while actively *hiding* content that is too dissimilar. Due to time constraints or even simplicity of use, users often rely on recommendations only to discover new interests, opinions or contents; in turn, these

A. Hameurlain et al. (Eds.): *Transactions on Large-Scale Data- and Knowledge-Centered Systems LIX*, LNCS 16240, pp. 1–31, 2026.
https://doi.org/10.1007/978-3-662-72449-1_1

Fig. 1. Example of diverse recommendations (Football dataset). Node colors represent communities as computed by a greedy modularity algorithm. The red square is the target node and red diamonds are diverse recommendations, i.e., from other communities. (Color figure online)

recommendation algorithms do not change the *exposure to information* of the user.

All these dynamics contribute to an environment that naturally facilitates the spread of *fake news*, as no counterpoints are readily available and the shock value maintains user attention; as well as the formation of *echo chambers* when users tend to aggregate around common interests, or *filter bubbles* when recommender systems are over tuned towards similarity-based recommendations. This is problematic as this can lead towards segregation of opinions and mass polarization.

Research to measure the full extent of the possible harmful consequences is ongoing [6,10,16,20,35]; such research is important, especially since society is slowly letting go of traditional news outlets in favor of news propagated through social media [6,37]. Algorithmically, ways to optimize diversity and reduce polarization and disagreement in social graphs have been proposed: either based on a global measure of polarization [7,31,43], optimizing exposure to a variety of opinions [30], or by assuming that reducing overall distances in the graphs will increase diversity [9].

Most of these approaches rely on reconfiguring the graph at a *global* level, thus assuming that one can make changes affecting any user in the graph. We believe that *local* optimization, i.e., from the point of view of a single user, may allow providing more personalized recommendations, minimize the global effect in the graph (indeed, other users might not desire diversity); in a word, increase effectiveness and chance of adoption. To achieve this, we need to provide complementary semantics to how to find new content in social networks. Such semantics can be based on maximizing a measure of *diversity* of the content that a given user is exposed to; indeed, if we consider that opinions are clustered into *communities* one way to see diversity is to promote the exposure of users to content from communities different from their own. An example is given in Fig. 1: when the node identified by the red square gets linked to the red diamond nodes—part of other communities —, it is directly exposed to more, diverse sources of information. Intuitively, it also is a way to increase the *entropy of the information exposure* from the node's point of view.

Contributions. In this paper, we present an approach to optimizing the *diversity* of *information exposure* in the context of social networks based on a general formulation of exposure as the distribution of opinions (represented as not necessarily distinct communities) a given node u is exposed to. This allows us to: (i) be flexible in the definition of the exposure function, and we take inspiration from well-studied approaches such as the *polarization* of opinions in the graph [31], but also from classic random-walk and distance-based metrics; (ii) going beyond polarization, to model diversity of exposure over more than 2 polarized opinions; and (iii) define diversity not only as targeting uniform distributions of opinions but *any* distribution of opinions – crucially allowing us to incorporate user interests in the optimization.

This article is an extension of the previously published conference paper [8]. We present a new formulation that broadens the previous framework to take into account multiple propagation channels, not only through social links. In particular, we consider the propagation of information through a graph of items (documents, videos, posts, etc.) that is linked to the social network of users. This new formulation requires new results with regards to the propagation model and also the computation of the gradients in both parts of the new hybrid graph. We validate this setting to the experimental study under two formulations: user-user and user-item recommendation.

Outline. We formulate the setting of opinions as representing partitions (or communities) in the graph, define the diversity of exposure as a function of the distribution of opinions, and link it with entropy-based measures in Sect. 3. We present the matrix formulation of the problem in Sect. 4 and provide a gradient descent algorithm, along with two reasonable greedy-based heuristics. We extend this setting to the user-item graph in Sect. 5. Finally, we compare our algorithms to algorithms in the state-of-the-art on a varied set of real-world graphs, covering ground truth and inferred communities along with non-discrete ones, in Sect. 6.

2 Related Work

The objective of social media companies is to maximise user attention and time spent on their platform. This was enabled by their recommender systems that recommend new content based on the likelihood that the content is liked by the users. The link recommendation problem has been extensively researched and there are various approaches to solving it using network embedding [18] and deep learning methods such as graph neural networks [23].

However, these provide recommendation based on similarity only and the diversity suffers from it. Some diversification methods have been proposed based on re-ranking algorithms [42], but the issue of diversity in recommender systems is still a growing focus in both academic and industry research.

Indeed, there are rising concerns about social networks. Research has focused on the effects and mechanisms of *filter bubbles*, *echo chambers*, misinformation spread and Polarization. Even if not unanimously agreed upon [27,35] it is considered that there are links between growing social fragmentation and recommender systems. In order to study those processes in social networks, an opinion dynamic model is used, the Friedkin-Johnsen model is the most widely adopted model. It uses an aggregation similar to that of the (personalized) PageRank algorithm [32] to model the opinion of a node as a function of its neighbors' opinions, it also differentiates between the personal innate opinion with the expressed opinion of a given person. Based on this model, some approaches have been proposed to minimize polarization [11,13,21] and disagreement [7,31,43]. While effective, these only capture online interaction through the lens of a single dimension of opinion ranging from 0 or -1 to 1. It does not allow for multiple poles of opinions, arguably a more realistic scenario.

Other solutions based on information cascades have also been proposed. Originally aimed at political or marketing campaigns in the context of influence maximization, these methods can also be extended to the goal of maximizing information spread. Matakos et. al [30] maximize the diversity of exposure to a seed set of items using independent cascade model.

There are also concerns about the fairness of recommender systems which can exacerbate some biases – such as popularity, selection or position [5,33]. These biases are undesirable and may contribute to filter bubbles. Hence fairness-aware approaches have gained increasing attention [15,29] along with purely diversity focused methods [22,41]. However these approaches tend to disregard social dynamics and can be contradicting actual user behavior as they are neither based on any opinion dynamics or cascade model.

Another approach is to maximize *serendipity*. Even though definitions of serendipity vary [24,44], they seem consistent on two main properties: interest and novelty. Content that is unknown (so novel if recommended) yet interesting to a user is considered serendipitous. However, concrete results in this area are quite sparse. Indeed, there is a difficulty of measuring serendipity. Current approaches rely on user inputs; these are quite rare commodities on online platforms as users usually dislike having to share personal thought on a given post – consider the number of likes/dislikes and comments compared to number of views on videos on e.g., YouTube. Moreover, datasets that provide such information are far and few between, making benchmarks and comparisons difficult.

Another complexity of social network is in its structure. Indeed, there are not only users but also contents generated by these users and then disseminated. Previous models – Friedkin-Johnsen opinion dynamics [7,31,36], information cascades [14,34] – only account for user-user interaction. This means that propagation through similar content (for instance, recommendation of similar items in video platforms) is only *implicitly* represented while ignoring explicit propagation and influence on the users' opinions.

Many of the previously mentioned algorithms are also based on a greedy structure which has optimality guarantees only when coupled with sub-modular objectives. In this work, we take inspiration from [7,31] which uses a matrix representation of the social network in order to use gradient descent algorithms enabled by the convex nature of the formulation of their problem and of the feasible set. Abebe et. al [1] show that interesting results can also be obtained in a non-convex environment. One strong advantage of such approach is the absence of needing to specify an explicit candidate set of recommendations, which saves us the issue of how to conceive it and the overall computational cost.

3   Preliminaries

We represent the social network as a connected directed graph $G = (V, E)$, where V is the set of nodes and $E \subseteq V \times V$ the set of edges, $n = |V|$, $m = |E|$. We use the standard definitions of the adjacency matrix A_G (with $A_{ij} > 0$ if $(i, j) \in E$ and 0 otherwise) and the degree matrix D_G (with $D_{ii} = \sum_j A_{ij}$). We denote by $\mathcal{N}_u$ the set of neighbor indices of u in G. When unambiguous the subscript G is omitted.

To represent the fact that the social network is organized in communities (possibly having different semantics), we define the notion of a *graph partition* $\mathcal{P}$:

Definition 1 (Graph Partition). *A graph partition of a graph G, $\mathcal{P}_G$, is a set of subsets of nodes $\mathcal{X}_i \subset V$:*

$$\mathcal{P}_G = \left\{ \mathcal{X}_i \mid \mathcal{X}_i \subset V, \mathcal{X}_i \neq \emptyset, \bigcup^i \mathcal{X}_i = V \right\}. \tag{1}$$

The set of partitions to which a node u belongs is denoted as:

$$\mathcal{P}_G(u) = \{ \mathcal{X}_i \mid u \in \mathcal{X}_i \}. \tag{2}$$

Note that this definition allows for overlapping partitions. So in this sense, the use of the term "partitions" is incorrect; however, we keep as such as most communities in graphs, either real or detected using algorithms, are disjoint.

In matrix form, we represent the partitions as a column-stochastic matrix P of dimension $p \times n$, with the property that $\sum_i P_{ij} = 1$, $\forall j$. The column encode the distribution of communities of opinions an individual node has.

3.1   Node Exposure

We define next the concept of *exposure to information* and give some examples of how it can be measured in a graph.

Definition 2 (Node Exposure). *The* node exposure *function is a function* $\mathcal{E} : V \to \mathbb{R}^n$, *representing the distribution of exposure of information from all nodes* $v \in V$; *moreover, we require that this is a distribution, i.e.,* $\mathbf{1}^\top \cdot \mathcal{E}(u) = 1$, $\forall u \in V$.

From node exposure, we can naturally extend it to a *partition exposure* function, $\mathcal{E}_P : V \to \mathbb{R}^p$, such that

$$\mathcal{E}(u \mid \mathcal{P}_G) = \mathcal{E}(u) \cdot P.$$

It is easy to check by the properties of $\mathcal{E}$ and $\mathcal{P}_G$ that $\mathcal{E}(u \mid \mathcal{P}_G)$ is also a distribution.

Exposure Functions. There exist various ways of defining the exposure function $\mathcal{E}$, and we present here some options.

Personalized Random Walk (PPR). This is the classic PageRank random walk process [32], rooted at a node u using transition weight matrix $W_u = \alpha D^{-1}A + (1-\alpha)(\mathbf{1}_n \otimes \mathbf{u})$, where $\mathbf{u}$ is the one-hot vector of u, and $\otimes$ is the Kronecker product. The exposure distribution is then the stationary distribution of this process:

$$\mathcal{E}(u) = W_u^\infty = (1-\alpha)\left(I - \alpha D^{-1}A\right)^{-1} \mathbf{u},$$

where $\mathbf{u}$ is the one-hot encoding of u.

Friedkin-Johnsen Model (F-J). This is a model of opinion dynamics in social networks [12], where the opinion of a node u is an aggregation of opinions of neighbors. It is equivalent to a PPR on the transpose graph, if we consider that the initial vector of opinions is $\mathbf{u}$:

$$\mathcal{E}(u) = W_u^\infty = (1-\alpha)\left(I - \alpha D^{\top^{-1}}A^\top\right)^{-1} \mathbf{u}.$$

Note that when $\alpha = 0.5$ and $D = I$, this is equivalent to the stationary distribution found in [7,31].

Breadth-First (BFS). This is a simple exposure model, where the exposure of a node u is inversely proportional to the depth at which each node v is found in the process of a BFS search.

$$\mathcal{E}(u) = \sum_i A^i$$

Example 1. Figure 2a exemplifies a small social network of 15 nodes. In this example we consider the point of view of node $u = 13$. The network is divided into 4 partitions as shown in Fig. 2b – in this case, generated using the Louvain modularity algorithm. We exemplify node 13 and the BFS exposure (cut-off at depth 2 for legibility reasons), shown in Fig. 2c. The resulting distribution over the 4 partitions is: $\mathcal{E}(13 \mid \mathcal{P}_G) = \{0.125, 0.125, 0, 0.750\}$. This is because in the search, the blue partition is reached once (node 3 at depth 2), the purple partition once (node 0 at depth 2), the green partition never, and the "home" yellow partition 4 times (nodes 8, and 12 at depth 1, and 12 and 14 at depth 2).

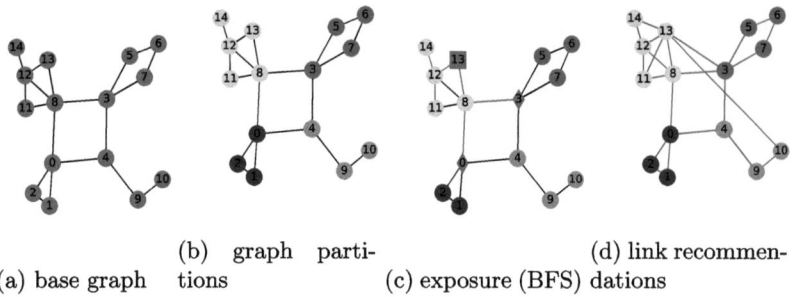

(a) base graph   (b) graph parti-
tions   (c) exposure (BFS)   (d) link recommendations

Fig. 2. Example social network: partitions, vision, and recommendations.

3.2   Diversity of Exposure

Given the probability distribution of the exposure function $\mathcal{E}$ we can now define a diversity measure function $\sigma\left(\mathcal{E}\left(u \mid \mathcal{P}_G\right)\right) \in \mathbb{R}$ which quantifies—given a node u in a graph G partitioned according to $\mathcal{P}_G$—how diverse u's exposure $\mathcal{E}(u \mid \mathcal{P}_G)$ is.

Intuitively, if we view the partitions in $\mathcal{P}_G$ as representing communities of opinion, the aim of diversity of exposure measure σ is to be as close as possible to a distribution in which all opinions / partitions are equally represented.

Entropy. A reasonable candidate measure is then **Shannon's entropy**—which is maximized when the distribution is equi-probable:

$$\sigma\left(\mathcal{E}\left(u \mid \mathcal{P}_G\right)\right) := \mathcal{H}\left(\mathcal{E}\left(u \mid \mathcal{P}_G\right)\right) = -\sum_{\mathcal{X}_i \in \mathcal{P}_G} p(\mathcal{X}_i) \log(p(\mathcal{X}_i)). \tag{3}$$

Entropy makes intuitive sense because of its aim to measure the amount of information in a message. Entropy-based measure have been used before to measure diversity [26] and fairness [33,39]. Moreover, [36] showed experimentally that the entropy can also quantify polarization in a graph.

Cross-entropy. Another way to look at the problem is that we are aiming to minimize the distribution distance to a reference desired distribution Q. This can occur when we would like to minimize the exposure to e.g., toxic partitions while keeping the other partitions diverse. This can be measured using the **KL divergence** between $\mathcal{E}$ and Q:

$$\sigma_Q\left(\mathcal{E}\left(u \mid \mathcal{P}_G\right)\right) := \mathcal{D}_{\mathrm{KL}}\left(Q \| \mathcal{E}\left(u \mid \mathcal{P}_G\right)\right) = \tag{4}$$

$$= -\sum_{\mathcal{X}_i \in \mathcal{P}_G} q(\mathcal{X}_i) \log\left(\frac{q(\mathcal{X}_i)}{p(\mathcal{X}_i)}\right).$$

When Q is the equi-probable distribution, Eq. (4) is equivalent to Eq. (3).

Example 2. We return to the example in Fig. 2. The BFS vision distribution is associated to an entropy value of 0.736. When one recommends using classic, triadic closure based, recommendation algorithms, node 13's community becomes denser (edges $13 - 14$ and $13 - 11$, in red in Fig. 2d), resulting in a more "biased" exposure: $\mathcal{E}(13 \mid \mathcal{P}_G) = \{0.1, 0.1, 0, 0.8\}$ and a smaller entropy value of 0.639. On the other hand, when recommending by optimizing entropy-based measures (edges $13 - 10$ and $13 - 3$, in green in Fig. 2d) we get a more balanced vision: $\mathcal{E}(13 \mid \mathcal{P}_G) = \{0.145, 0.285, 0.285, 0.285\}$ and a higher entropy of 1.351. Moreover, we can see that this is achieved by linking 13's original community, yellow, with the green and blue communities.

4   User-Based Diversity: Problem and Algorithms

The studied problem can then be stated as follows:

Problem 1 (Single-Node Exposure Diversity Maximization). Given a graph $G = (V, E)$, a node $u \in V$, a partition $\mathcal{P}_G$, an exposure function $\mathcal{E}(u)$ *maximize* the diversity measure $\sigma(\mathcal{E}(u \mid \mathcal{P}_G))$ for a fixed number k of *edge additions* to the graph starting at u.

We know that both entropy and KL divergence are maximized at the equiprobable or the Q distributions, respectively. Assuming that we have a set of *feasible graphs* $\mathcal{G}$ (defined later), the optimization problem can be stated as the minimization of the distance between the exposure distribution and these distributions:

$$\arg\min_{G \in \mathcal{G}} ||\mathbf{d}||^2 = ||\mathcal{E}_G(u \mid \mathcal{P}) - \mathbf{q}||^2 = ||P \cdot \mathcal{E}_G(u) \cdot \mathbf{u} - \mathbf{q}||^2, \tag{5}$$

where $\mathcal{E}_G$ denotes the exposure function computed on the graph G, $\mathbf{u}$ the one-hot encoding of u, and $\mathbf{q}$ is the desired target distribution.

Note that the formulation in Eq. 5 is similar to the formulation for minimizing polarization in a Friedkin-Johnsen model as the distance between the vector of opinions and the average opinion in the graph, as in [2,7,43]. We claim that our formulation is more general, as minimizing the distance to the average opinion risks decreasing polarization but increasing extreme opinions, e.g., when the average opinion is extreme (note that this can also occur when the average opinion is 0-centered *a posteriori*, as claimed in [43]).

Properties of the Objective Function. The objective is *non-monotone*. This means that the greedy algorithm is not guaranteed to find the optimal solution; indeed, as we will show in the experimental results, there comes a point of diminishing returns when adding edges to the graph actually *decreases* the diversity of exposure.

Moreover, the objective is not always convex. On the one hand, for the PPR and F-J functions, the objective is *convex*, as it has the same form as the polarization minimization in [31]. On the other hand, for the BFS function – and when

expressing the exposure function as a sum of powers of the adjacency matrix –
one can see that, from depth $d = 3$ on, the objective is not convex due to at
least the term A^3.

4.1 Convex Set of Feasible Graphs

For all exposure functions, we can apply gradient descent methods. Consider the
following set of graphs:

$$\mathcal{G}_u(A) = \{G' \mid A'_{ij} > 0, \forall (i, j) \in E(G), \tag{6}$$
$$A'_{ij} = 0, \forall (i, j) \notin E(G), i \neq u,$$
$$\sum_j A'_{ij} = 1, \forall i \in V(G)\},$$

i.e., all graphs that have their edge weights normalized, and for which $D = I$.
Moreover, we allow only graphs that change only in row u (adding new links
from u to other nodes). It is straightforward to show that this set is convex: a
linear combination of row-stochastic matrices is also row-stochastic; it is also the
set defined in [7,31], restricted to u:

Proposition 1. *The set of feasible graphs $\mathcal{G}_u(A)$ is convex.*

By Eqs. 5 and 6 we can adapt gradient descent algorithms for Problem 1. For
this, we have to define the gradient of the objective function, and the projection
step.

Gradient Step. The derivative of the objective depends on $\frac{\partial \mathcal{E}}{\partial A}$ in the following
way:

Proposition 2. *Assuming that $\mathcal{E}$ is a derivable function of the adjacency matrix
A, the gradient of the objective function is given by:*

$$\frac{\partial \mathcal{O}}{\partial A} = 2\mathbf{d} \left(\mathbf{u}^\top \otimes P\right) \frac{\partial \mathcal{E}}{\partial A}. \tag{7}$$

Proof (sketch). Follows from writing $\mathcal{O}$ in algebraic form, $\mathcal{O} = \mathbf{d}^\top \mathbf{d}$ and then
using the product rule of matrix calculus. [4]

This general form allows us to "plug-in" any exposure function and its derivative (or simply provide a numerical approximation thereof). For PPR and F-J
exposure:

Proposition 3. *The gradient of the objective function for the **PPR** and **F-J**
exposure functions is given by:*

$$\frac{\partial \mathcal{E}}{\partial A} = -\alpha(1 - \alpha)(W_\infty^{-\top} \otimes W_\infty^{-1}),$$

where $W_\infty = I - \alpha A$ for PPR, and $W_\infty = I - \alpha A^\top$ for F-J.

Proof (sketch). We use the facts that:

$$dW_\infty = -\alpha dA,$$
$$d\mathcal{E} = -(1-\alpha)(W_\infty^{-1}dW_\infty W_\infty^{-1})$$
$$= (1-\alpha)(W_\infty^{-\top} \otimes W_\infty^{-1})dW_\infty.$$

The final formula follows.

We can express the BFS exposure function as a sum of powers of A, limited to a depth of 3 for practical reasons:

$$\mathcal{E}^{(\text{BFS})} = \sum_{i=1}^{3} A^i = A(I - A^3)(I - A)^{-1}.$$

Proposition 4. *The gradient of the objective function for the **BFS** exposure function at depth 3 is given by:*

$$\frac{\partial \mathcal{E}}{\partial A} = I + (I \otimes A + A^T \otimes I)+$$
$$+ (A^{T^2} \otimes I + A^T \otimes A + I \otimes A^2).$$

Proof (sketch). This follows from developing the $\mathcal{E} = \sum_{i=1}^{3} A^i$ as $d\mathcal{E} = dA + dA^2 + dA^3$ and applying the rules of derivatives of powers of matrices.

Note that we could straightforwardly expand this to higher powers, but it would require both more involved formulas and would cost more computation time.

Once we have the matrix-function gradients, we can directly apply projected gradient descent algorithms, alternating gradient descent and projection in the feasible set (i.e., a modified graph). Algorithm 1 details the steps.

Algorithm 1: DESCENTDIVERSE

Data: graph G, target node u, exposure function $\mathcal{E}$, graph partition $\mathcal{P}_G$, target distribution $\mathbf{q}$, number of edges k, rate η

1 $G^{(1)} \leftarrow G$, $A^{(1)} \leftarrow A$;
2 **for** $i \in \{1, \ldots, k\}$ **do**
3   compute $\mathcal{O}_i = ||\mathcal{E}_{G^{(i)}}(u \mid \mathcal{P}) - \mathbf{q}||^2$;
4   gradient: $\Delta A^{(i)} = \frac{\partial \mathcal{O}_i}{\partial A}$ according to Eq. 7;
5   descent: $A' \leftarrow A^{(i)} - \eta \Delta A^{(i)}$;
6   projection: $\{v_i, A'_u\} \leftarrow$ PROJECT(u, A'_u);
7   add to graph: $G^{(i+1)} \leftarrow G^{(i)} + (u, v_i)$, $A^{(i+1)} = A$, $A_u^{(i+1)} = A'_u$;
8 **end**
9 **return** $\{v_1, \ldots, v_k\}$

Projected gradient descent algorithms are used in online convex optimization, and are know to have a regret of $O(\sqrt{k})$ for $\eta = O(1/\sqrt{k})$ [45], *if* the function $\mathcal{O}$ is convex. For us, this is not generally the case, but we show that DESCENTDIVERSE is effective in practice.

Algorithm 2: PROJECT

Data: target node u, set of neighbor indices $\mathcal{N}_u$, u's row in the modified
adjacency matrix A'_u

1 get the candidate set of nodes which have positive values:
$\mathcal{C}(u) = \{v \mid v \notin \mathcal{N}_u, A'_{uv} > 0\}$;
2 get $v = \arg\max_{v \in \mathcal{C}(u)} A'_{uv}$;
3 set all other edges to 0: $A'_{uj} = \mathbf{0}, \forall j \notin \mathcal{N}_u \cup \{v\}$;
4 re-normalize the edge weights: $A'_{ui} = 1/|\mathcal{N}_u \cup \{v\}|$;
5 **return** $\{v, A'_u\}$;

Projection Step. The projection step ensures that we only add edges from the candidate node u, and that the edge weights are normalized, thus keeping the modified graph in the feasible set $\mathcal{G}_u$. The steps are detailed in Algorithm 2.

Optimizing the Gradient Computation. The gradient computation is the main bottleneck of Algorithm 1. However, we can note that the derivative of the objective function depends on the whole adjacency matrix A but we only care about the final distribution for node u. The Kronecker product $(\mathbf{u}^\top \otimes P)$ is in fact a (very) sparse matrix, only having P for the coordinates corresponding to u. Moreover, we can simplify each Kronecker product by identifying all indices corresponding to the target node u:

– $(\mathbf{u}^\top \otimes P)$ simply becomes P at indices $[u, u]$
– for **PPR** and **F-J**, we only need the square sub-matrix of size n at $[un, un]$ in the Kronecker product $Q^{-\top} \otimes Q^{-1}$ which becomes $Q_{u,u}^{-\top} Q^{-1}$, hence the local gradient at u becomes (abusing notation):

$$\frac{\partial \mathcal{E}}{\partial A}[u] = -\alpha(1 - \alpha)(I - \alpha A)_{u,u}^{-T}(I - \alpha A)^{-1};$$

– for **BFS**, the Kronecker products simplify naturally if we only consider the local gradient at u: (i) $I \otimes A$ becomes A, (ii) $I \otimes A^2$ becomes A^2, (iii) $A^T \otimes I$ and $A^T \otimes A$ disappear in the sum because A's diagonal values are zeros, and (iv) $A^{T^2} \otimes I$ becomes $A^{T^2}_{u,u}$, leading to:

$$\frac{\partial \mathcal{E}}{\partial A}[u] = I + A + A^2 + A^{T^2}_{u,u}.$$

In order to avoid computing the inverse, we use the Biconjugate gradient stabilized method (BiCGSTAB) to approximate it. This approximation is given by solving for $\mathbf{x}$:

$$W\mathbf{x} = P^+\mathbf{q} + \mathbf{u}, \tag{8}$$

where W is either $I - \alpha A$ or $I - \alpha A^T$, P^+ is the Moore-Penrose inverse of P, and $\mathbf{q}$ the target distribution. The resulting solution corresponds directly to the exposure of node u. This change of shape, from matrix $n \times n$ to a vector of length n, entails minimal changes to Algorithm 1, where the projected gradient descent is performed directly on the node's exposure. No changes are need in the the projection step, Algorithm 2.

Computational Complexity. For the DESCENTDIVERSE algorithm, the first computation of the inverse matrix for the PPR and F-J exposure is unavoidable and has a complexity of $O(n^3)$, but the subsequent updates have a complexity of $O(n^2)$, as we only need to update the row and column of the target node u. For the BFS exposure function at depth 3, the initial complexity is also $O(n^3)$, but subsequent computations are $\mathcal{O}(n^2)$, as we only need to update the row and column of the target node u and the candidate node c. The overall complexity is thus $O(n^3 + kn^2)$.

The overall complexity of solving the linear equation system in Eq. 8 using the BiCGSTAB iterative method depends on the size of the system, the density of the graph and the convergence rate which depends on the condition number of $I - \alpha A$. We used the default and standard tolerance of 10^{-6} meaning that the residual error is lower than 10^{-6}. Hence, when using the BiCGSTAB method, the complexity of the gradient computation depends on the sparsity of the networks is $O(kRm)$, where R is the number of rounds to convergence.

4.2   Candidate Set of Edges

Another option, but which has a high computational cost, is to evaluate the objective over a set of *candidate edges* $\mathcal{C}$, which is potentially as large as all nodes $v \notin \mathcal{N}_u$, where $\mathcal{N}_u$ is the set of neighbors of u.

Under this formulation, Problem 1 can be stated as maximizing over all possible configurations of k edges in $\mathcal{C}$. Our objective function is not general monotone or sub-modular, so we rely on heuristics which add one recommendation at a time. We present here two alternatives to solve this problem: a "partition boosting" algorithm and a greedy algorithm.

Partition Boosting. This algorithm uses the intuition that, at any step, the best edge to add is one to the partition that is "farthest" from the objective $\mathbf{q}$, but only if it is under-represented[1]. Stated otherwise, we choose some partition from the set $\{\mathcal{X} \mid d_i < 0, \mathcal{X} \in \mathcal{P}_G\}$. We also assume that the set of candidates is the union of candidates per partition, i.e., $\mathcal{C} = \cup_i \mathcal{C}(\mathcal{X}_i)$.

Once the partition is chosen, we add the edge in $\mathcal{C}(\mathcal{X}_i)$ that maximizes the objective, and repeat the process until k edges are added, as detailed in Algorithm 3.

There are several ways to choose the partition in CHOOSEPARTITION, and we present three alternatives that we have evaluated experimentally:

1. *the most under-represented partition:* $\mathcal{X}_i = \arg\min_i d_i$;
2. *draw randomly from the under-represented partitions:* $\mathcal{X}_i \sim \mathcal{U}(\{\mathcal{X} \mid d_i < 0\})$;
3. *draw from a soft-max distribution:* $\mathcal{X}_i \sim \mathrm{SoftMax}(\mathbf{d})$.

---

[1] An interesting option would be to also allow removing edges from over-represented partitions, but this has obvious drawbacks in practice.

Algorithm 3: PARTITIONBOOSTINGDIVERSE

Data: graph G, target node u, exposure function $\mathcal{E}$, graph partition $\mathcal{P}_G$, target
distribution $\mathbf{q}$, number of edges k, candidates $\mathcal{C} \subseteq V$

1 **for** $i \in \{1, \ldots, k\}$ **do**
2 | $\mathcal{X}_i = $ CHOOSEPARTITION$(\mathbf{d})$;
3 | $v_i = \arg\min_{v \in \mathcal{C}(\mathcal{X}_i)} \|\mathcal{E}_{G+(u,v)}(u \mid \mathcal{P}_G) - \mathbf{q}\|^2$;
4 | add edge (u, v_i) to G: $G \leftarrow G + (u, v)$;
5 **end**
6 **return** $\{v_1, \ldots, v_k\}$

Greedy Algorithm. The objective function that we have defined, including the entropy and the KL-divergence, is not monotone. There is hence no hope of having an approximation algorithm using the greedy algorithm *in general*. However, it remains an interesting heuristic: at each step, we add the edge that maximizes the marginal gain in the objective function, as detailed in Algorithm 4.

Algorithm 4: GREEDYDIVERSE

Data: graph G, target node u, exposure function $\mathcal{E}$, graph partition $\mathcal{P}_G$, target
distribution $\mathbf{q}$, number of edges k, candidates $\mathcal{C} \subseteq V$

1 **for** $i \in \{1, \ldots, k\}$ **do**
2 | $v_i = \arg\min_{v \in \mathcal{C}} \|\mathcal{E}_{G+(u,v)}(u \mid \mathcal{P}_G) - \mathbf{q}\|^2$;
3 | add edge (u, v_i) to G: $G \leftarrow G + (u, v)$;
4 **end**
5 **return** $\{v_1, \ldots, v_k\}$

Optimizing the Computation. Instead of computing the full objective function at each step, we can compute incrementally the changes in the exposure function.

For **BFS** exposure, instead of computing powers of the matrix A, we can simply update the matrix A considering only the node u and the index of the candidate that has been added in the previous step, c. Once the initial $\mathcal{E}$ has been computed, we can accurately estimate the changes in the exposure function by taking advantage of the following intuitions: (i) at hop 2 we have to add the neighbors of u to c and vice versa, by transitivity, i.e., add the vector A_u to the column A_c and the column $A_c^\top$ to the row A_u; and (ii) at hop 3 we have to add the neighbors of the neighbors of u to the neighbors of c and vice versa.

For **PPR** and **F-J** exposure, we can incrementally update the inverses using the Sherman-Morrison formula.

Computational Complexity. For the PARTITIONBOOSTINGDIVERSE and GREEDYDIVERSE algorithms, the complexity is $\mathcal{O}(kn^2)$, as we need to compute the objective function for each candidate node v, by using the optimizations that have been described above. Moreover, we have to do it for each of the candidates at each step of the algorithm. However, the PARTITIONBOOSTINGDIVERSE algorithm only needs to compute for at most $\mathcal{C}_{\max} := \max_i |\mathcal{C}(\mathcal{X}_i)|$ candidates. Hence the complexity is $O(n^3 + k\mathcal{C}_{\max}n^2)$ for PARTITIONBOOSTINGDIVERSE and $O(n^3 + k\mathcal{C}n^2)$ for GREEDYDIVERSE.

5   Content-Based Extension: User-Item Graph

The previous model captures the framework of a user-user (social) graph, where we only have access to user information.

We extend our social setting to user-item graphs to include items created by users, which can be consumed (liked or shared) by other users. In this framework, we define $G = (U \times C) = (E_G = \{E_U, E_C, E_T\}, V_G = \{V_U, V_C\})$, where G represents the entire graph, U the user graph, and C the content (or item) graph. Edges are directed and categorized as E_U (between users), E_C (between contents), or $E_T = E_{V_U \cap V_C}$ (connecting users and contents).

Propagation Model. We keep the same two base assumptions as in the previous model. First, we still consider that information travels along all edges in the above hybrid graph, even if the relationship semantics for a link between users, between items, or between both might differ. This means that there are more pathways for the propagation of information. Secondly, we assume that the partitions are separate: the user partitions have the same meaning as the one discussed before, while the items are clustered in partitions only on the item-item graph (via similarity, for instance).

Problem Statements. The previously defined problem now has two variants: we optimize either along user edges or item. In other words, we recommend user nodes or item nodes:

Problem 2 (Single-Node User Exposure Diversity Maximization). Given a graph $G = (U \times C)$, a node $u \in V$, a partition $\mathcal{P}_U$, an exposure function $\mathcal{E}(u)$ *maximize* the diversity measure $\sigma(\mathcal{E}(u \mid \mathcal{P}_G))$ for a fixed number k of *edge additions in U* to the graph starting at u.

Problem 3 (Single-Node Item Exposure Diversity Maximization). Given a graph $G = (U \times c)$, a node $u \in V$, a partition $\mathcal{P}_U$, an exposure function $\mathcal{E}(u)$ *maximize* the diversity measure $\sigma(\mathcal{E}(u \mid \mathcal{P}_G))$ for a fixed number k of *edge additions in I* to the graph starting at u.

For the algorithms in Sect. 4 to work we have to redefine how exposure in the graph is estimated, and then we have to adapt the derivation step. This is because the exposure is now twofold: a given user is influenced by their connection (user neighbors) or by the content they consumed (item neighbors).

We model the graph $G = (U \times C)$ as a block matrix and use the same objective function:

$$G = \begin{bmatrix} U & T \\ T^\top & C \end{bmatrix}; P = \begin{bmatrix} P_U & 0 \\ 0 & P_C \end{bmatrix}; \mathbf{u} = \begin{bmatrix} \mathbf{u} \\ T^\top \mathbf{u} \end{bmatrix}; Q = \begin{bmatrix} \mathbf{q}_U \\ \mathbf{q}_C \end{bmatrix}$$

$$\arg \min_{G \in \mathcal{G}} ||\mathbf{d}||^2 = ||\mathcal{E}_G(u \mid \mathcal{P}) - Q||^2 = ||P \cdot \mathcal{E}_G(u) \cdot \mathbf{u} - Q||^2,$$

This model does not require complete knowledge of the social network. If links between contents (or users) are unavailable, meaning C (or U) is empty, they can be substituted with the identity matrix. While the general case assumes two distinct partitions based on different semantics, either partition can be unknown (represented as an empty matrix) without affecting the optimization of the other. In the most general scenario, both partitions can coexist simultaneously.

We now state the new exposure functions for the user-item graph:

Proposition 5 (PPR Exposure). *Given the block matrix formulation of a graph $G = (U \times C)$ such that C has an invertible Schur complement S in G, then the PPR exposure is given by:*

$$\mathcal{E}_G^{PPR} = (1 - \alpha) \begin{bmatrix} S^{-1} \\ -C^{-1}T^\top S^{-1} \end{bmatrix} \tag{9}$$

with $S = \left(I - \alpha U - \alpha^2 T (I - \alpha C)^{-1} T^\top \right)$

Proposition 6 (F-J Exposure). *Given the block matrix formulation of a graph $G = (U \times C)$ such that C has an invertible Schur complement S in G, then the F-J exposure is given by:*

$$\mathcal{E}_G^{FJ} = (1 - \alpha) \begin{bmatrix} S^{-1} \\ -C^{-\top}T^\top S^{-1} \end{bmatrix} \tag{10}$$

with $S = \left(I - \alpha U^\top - \alpha^2 T (I - \alpha C^\top)^{-1} T^\top \right)$

Proposition 7 (BFS Exposure). *For any block matrix formulation of a graph $G = (U \times C)$, the BFS (limited to a depth of 3) exposure is given by:*

$$\mathcal{E}_G^{BFS} = \begin{bmatrix} U^3 + U^2 + U + UTT^\top + TT^\top U + TCT^\top + TT^\top \\ T^\top U^2 + T^\top U + T^\top TT^\top + CT^\top U + CT^\top + C^2 T^\top + T^\top \end{bmatrix} \tag{11}$$

Example 3. Figure 3a illustrates a social network where the user graph (colored nodes) is combined with a content graph (gray nodes). Each color represents a partition, consistent with the semantics in Fig. 2b. Black edges between users indicate relationships such as friendship, subscription, or following, while gray edges between contents signify similarity or references. Dotted gray edges connecting users and contents represent user interactions with specific items.

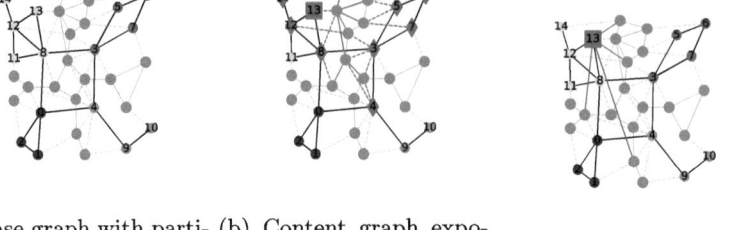

(a) base graph with partitions

(b) Content graph exposure (BFS)

(c) link recommendations

Fig. 3. Example social network and recommendations with the user-item graph added.

Figure 3b highlights the influence of the content graph (gray nodes) on the exposure of a target node. When the content graph is incorporated, the target node must first be projected into the content graph, exposure computed, and then projected back into the user graph. For instance, a BFS with a depth cutoff of 2 may return nodes at distances ranging from 2 to 4. Using node 13 as an example, the BFS exposure in the content graph (cut off at depth 2) differs significantly from the exposure in the user graph (colored nodes) shown in Fig. 2c. The new distribution across the four partitions is $\mathcal{E}(13 \mid \mathcal{P}_G) = \{0, 0.62, 0.14, 0.24\}$.

As a result, the diverse recommendations (green lines) in Fig. 2d become less diverse because node 13 is already highly exposed to the blue partition and even to the green partition, while remaining completely unexposed to the purple partition. Finally, Fig. 3c demonstrates potential diverse content recommendations (green lines) in this context. These new links expose node 13 to both the previously unseen purple partition and the green partition.

Derivations for User-Item Graphs. The derivation formulas for the GREEDYDI-VERSE algorithm remain consistent with 7, but $\partial \mathcal{E}$ varies depending on whether the optimization is over users or contents, i.e., whether U or T is modified. Furthermore, PPR and F-J exposures allow for distinct propagation processes by using separate α parameters for each. For simplicity, we assume a single value of α for U, C, and T in the following.

Proposition 8 (PPR Exposure Derivative). *For the PPR exposure function, the derivatives corresponding to the two problems are given by:*

$$\frac{\partial \mathcal{E}_G^{PPR}}{\partial U} = -\alpha(1-\alpha) \left[\begin{matrix} \left(S_{u,u}^{-\top} S^{-1} \right) \\ \left(C^{-1} T^{\top} \right) \left(S_{u,u}^{-\top} S^{-1} \right) \end{matrix} \right] \tag{12}$$

$$\frac{\partial \mathcal{E}_G^{PPR}}{\partial T} = (1-\alpha) \left[\begin{matrix} -\alpha^2 \left(S_{[u,u]}^{-\top} S^{-1} \right) \left(T(I - \alpha C)^{-1} + (T(I - \alpha C)^{-\top}) \right) \\ -\alpha^2 \left(\left(C^{-1} T^{\top} \right) \left(S_{[u,u]}^{-\top} S^{-1} \right) \left(T(I - \alpha C)^{-1} + (T(I - \alpha C)^{-\top}) \right) \right) + S_{[u,u]}^{-\top} C^{-1} \end{matrix} \right] \tag{13}$$

Proposition 9 (F-J Exposure Derivative). *For the Friedkin-Johnsen expo-sure function, the derivatives corresponding to the two problems are given by:*

$$\frac{\partial \mathcal{E}_G^{FJ}}{\partial U}[u] = -\alpha(1-\alpha) \left[(C^{-\top} T^{\top}) \begin{matrix} S_{u,u}^{-1} S^{-\top} \\ (S_{u,u}^{-1} S^{-\top}) \end{matrix} \right] \tag{14}$$

$$\frac{\partial \mathcal{E}_G^{FJ}}{\partial T}[u] = (1-\alpha) \left[\begin{matrix} -\alpha^2 \left(S_{[u,u]}^{-\top} S^{-1} \right) \left(T(I-\alpha C^{\top})^{-1} + (T(I-\alpha C^{\top})^{-\top}) \right) \\ -\alpha^2 \left((C^{-\top} T^{\top}) \left(S_{[u,u]}^{-\top} S^{-1} \right) \left(T(I-\alpha C^{\top})^{-1} + (T(I-\alpha C^{\top})^{-\top}) \right) \right) + S_{[u,u]}^{-\top} C^{-\top} \end{matrix} \right] \tag{15}$$

Proposition 10 (BFS Exposure Derivative). *For the BFS exposure func-tion, the derivatives corresponding to the two problems are given by:*

$$\frac{\partial \mathcal{E}_G^{BFS}}{\partial U}[u] = \left[\begin{matrix} U_{u,u}^{T^2} I + U_{u,u}^{\top} U + U^2 + U + U_{u,u}^{\top} I + T T_{u,u}^{\top} I + T T^{\top} \\ T^{\top} U + U_{i,i}^{\top} T^{\top} + T^{\top} + C T^{\top} \end{matrix} \right] \tag{16}$$

$$\frac{\partial \mathcal{E}_G^{BFS}}{\partial T}[u] = \left[\begin{matrix} (U T_{u,u} I) + (T_{u,u} U) + (T_{u,u} U^{\top}) + (U^{\top} T_{u,u} I) + T C + T C^{\top} + T \\ U_{u,u}^{2^{\top}} I + U_{u,u}^{\top} I + T T_{u,u}^{\top} I + T^{\top} T + U_{u,u}^{\top} C + C + C^2 \end{matrix} \right] \tag{17}$$

Algorithms. In this extended setting, our previous algorithms can be adapted with minimal changes. The projected gradient descent algorithm remains the same, utilizing the updated formulas for the objective function and gradient. The projection step depends on the optimization target: if optimizing over users (U), the projection is applied to the user matrix (U); if optimizing over contents, the projection is applied to the transition matrix (T), as links are created between user nodes and content nodes.

Baseline algorithms also remain unchanged, requiring only a modification in the candidate selection process (users or items). The primary computational challenge lies in matrix inversion. Specifically, we use Eq. 8 with $W = S$, which is the only matrix inverse that needs to be recomputed at each step. Other inverses, such as C^{-1} and $(I - \alpha C)^{-1}$, can be precomputed. Importantly, we do not need to invert the entire adjacency matrix of the graph; only subsets of this larger matrix are required. Consequently, the overall complexity of the DESCENTDIVERSE algorithm in the user-item graph setting is $O(x^3 + kx^2)$, where $x = \max(|U|, |C|)$. Typically, $|C| > |U|$, as there are generally more content items than users (e.g., on platforms such as X, the number of posts exceeds the number of users).

6   Experiments

6.1   Experimental Setup

Implementation. We implemented our experiments in Python using the `NetworkX` package for graph manipulation and the computational packages numpy and scipy.sparse. The GREEDYDIVERSE and PARTITIONBOOSTINGDI-VERSE algorithms support multiprocessing for faster running time[2].

Datasets We evaluate our approach on several real graph datasets, described in Table 1.

Table 1. Dataset details. For the dataset marked with †, the communities represent the number of node features, and for datasets marked ‡ the number of communities found by the Louvain algorithm.

Dataset	Nodes	Edges	Existing	#Comm.
Polbooks	105	441	✓	3
Football [17]	115	613	✓	12
Movielens [19]	1,680	264,718	✗	19†
Movielens User-item [19]	(items) 943	65,419	✗	5‡
	(users) 1,680	264,718		19†
Reddit [25]	34,671	123,570	✗	54‡
Facebook [38]	63,392	816,886	✗	74‡
Amazon [40]	334,863	925,872	✗	230‡

The *MovieLens* dataset is built from the Movielens100k dataset, in which we consider only movies. Each movie has a distribution over 19 movie genres. We assign an edge between two movies if they share more than two thirds of their movie genres. After this linking process, each node's genre vector is updated as the average of all genres of its neighbors. Using these vectors we can then build the partition matrix P, where each row corresponds to a genre and each column to the genre vector of a node. This allows us to exemplify the functioning of our algorithms on *non-discrete partition*.

The *MovieLens User-item* dataset is also derived from the Movielens100k dataset. In this case, movies are partitioned based on their genres, serving as content, while users act as reviewers. Links between reviewers are established using cosine similarity of their features, and communities are then inferred using the Louvain algorithm. This results in two distinct graphs: one representing movies and their genres, and the other representing users and their similarities. These graphs are connected through the reviews linking users to movies.

The *Football* and *Polbooks* datasets have ground-truth communities, respectively the football conferences and the political alignment of the books. Additionally, for all datasets without ground truth we can infer a partition using a community detection algorithm, we used Louvain's algorithm for greedy modularity [3]. For all other datasets, we use the Louvain community detection algorithm to infer the partitions.

---

[2] https://github.com/Jonathan-COLIN/Optimizing-Diverse-Information-Exposure-in-Social-Graphs.

Target Node Selection. The optimization process depends greatly on the initial situation of the target node. Indeed, high degree nodes have very different exposures compared to very low degree nodes. As such, in the following, target nodes are selected according to their degree. We batch nodes into three groups of high, medium and low degree and select an equal amount of nodes among these three groups.

Generating Non-uniform Target Distributions. The target distribution q is the main factor in how the exposure is optimized. Indeed, a standard uniform distribution would lead to an equal representation of each partition and a unit vector used as a distribution would lead to only promoting a single partition. To show the power of *non-uniform target distributions*, we can adapt q to take user interests into account. We assume that for all users, their initial exposure before the optimization process represents their own interest. From the initial exposure distribution, we scale this distribution by adding the uniform distribution and a normal distribution centered at u's partition. This allows us to promote the partitions that are closer to the user's initial interest and reduce the probability of partitions that are further away. We do this for all datasets.

Targeting non-uniform partitions rather than uniform partitions allows us to present recommendation that are less "extreme" in their diversity, and which would allow the user to be gently nudged towards a more diverse exposure. Furthermore, if a given partition is recognized as harmful, we can improve the target distribution by setting the corresponding value to a very small value.

Baselines. We compare our approach to the following baseline algorithms: random assignment of edges (RANDOM), triadic closure (recommending based on common neighbors, TRIADICCLOSURE), polarization and disagreement minimization algorithms in the state-of-the-art, and a global and local diameter minimization algorithm. We compare the baselines with the GREEDY, PARTITIONBOOSTING and DESCENT algorithms.

We used the Polarization and Disagreement minimization algorithm presented in [43], in particular their simpleGreedy method. Given a candidate set of edges, it iteratively select the most promising one with regards to a standard measure of Polarization and Disagreement. Following their implementation we initialize our opinions uniformly between -1 and 1. Since this is a global algorithm, operating on the entire graph, we constrain the candidate set to only include edges starting from the parameter node u. Additionally, we also compare our approach to a graph diameter minimization algorithm as presented in [9]. Their global method identifies center nodes of the graph and connects them together, we adapted it to always include the target node in the set of center nodes. Their local method solves a decision problem using linear programming in order to identify nodes to add. We identify these algorithms by GLOBALDIAMETERREDUCTION and SINGLESOURCEDIAMETERREDUCTION respectively.

All experiments choose 30 target nodes, equally distributed among high-degree, medium-degree and low-degree nodes in their respective graph. We set $\alpha = 0.05$ and $\eta = 0.1$.

6.2   Results

Ground-Truth Communities. We start by presenting the results of the graphs for which we have ground-truth communities, *Polbooks* and *Football*, for both uniform and preferential target distributions, seen in Fig. 4. Generally, we notice that DESCENTDIVERSE manages to achieve its objective faster than other approaches. Moreover, as expected, TRIADICCLOSURE is among the least effective approaches. Interestingly, RANDOM is quite competitive (and always better than TRADICCLOSURE), which can be attributed to the relatively small sizes of the graphs. In a number of cases, the PARTITIONBOOSTINGDIVERSE does not manage to optimize past a certain point; indeed, once it can no longer find a partition to boost (it looks only for under-represented ones), it is liable to get "stuck" for the rest of the run. On the other hand, the diameter reduction approaches of [9], especially SINGLEDIAMETERREDUCTION, are very competitive, even for preferential distributions.

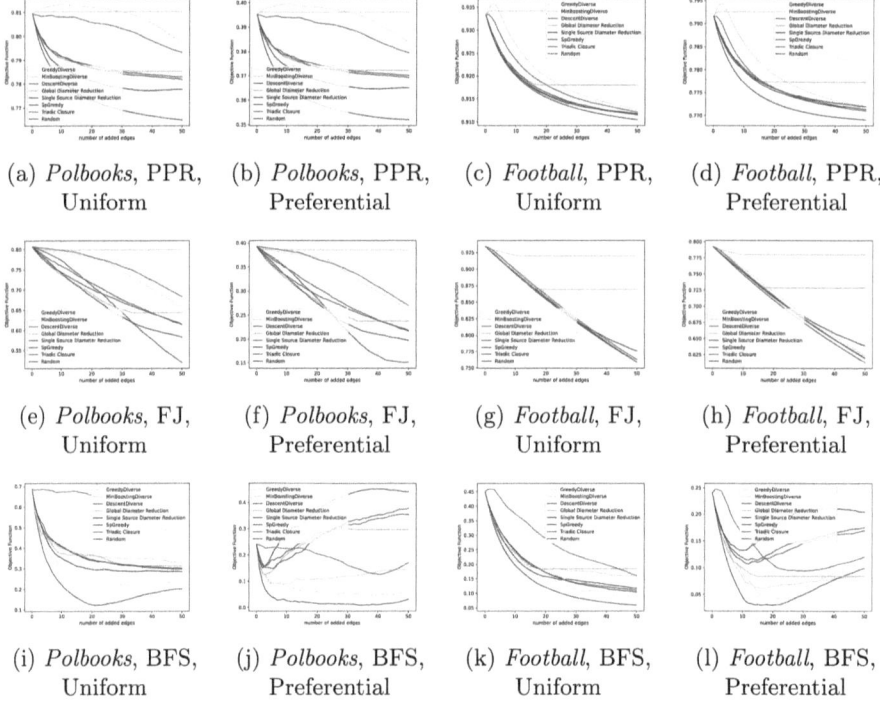

(a) *Polbooks*, PPR,   (b) *Polbooks*, PPR,   (c) *Football*, PPR,   (d) *Football*, PPR,
     Uniform   Preferential   Uniform   Preferential

(e) *Polbooks*, FJ,   (f) *Polbooks*, FJ,   (g) *Football*, FJ,   (h) *Football*, FJ,
     Uniform   Preferential   Uniform   Preferential

(i) *Polbooks*, BFS,   (j) *Polbooks*, BFS,   (k) *Football*, BFS,   (l) *Football*, BFS,
     Uniform   Preferential   Uniform   Preferential

Fig. 4. Objective vs. number of edge additions, ground truth communities.

One interesting, and potentially important, finding can be seen in the results on the BFS exposure function. One can notice that there exists a point in the process where adding new edges becomes counter-productive. This makes intuitive sense, seeing that the objectives are non-monotonous. Moreover, it means that one cannot indiscriminately add edges to maximize diversity, but that there is an optimal number of edges, which will depend on the graph and the node itself.

Inferred Communities. Where no ground truth is available, we generated partitions using a community detection algorithm. The results comparing RANDOM and TRIADICCLOSURE to DESCENTDIVERSE are shown in Fig. 5. The results are similar to those on graph having ground truth communities, with DESCENTDI-VERSE being the most effective at the end of the recommendation process (even if, in some cases, it is not the best at the start). Interestingly, recommending in the "classic" approach, TRIADICCLOSURE, is almost always worse than RANDOM. This makes some intuitive sense as recommending based on common neighbors would get the user "stuck" in the same community, whereas recommending random links at least gives a chance of escaping their current community and thus increasing the diversity of exposure.

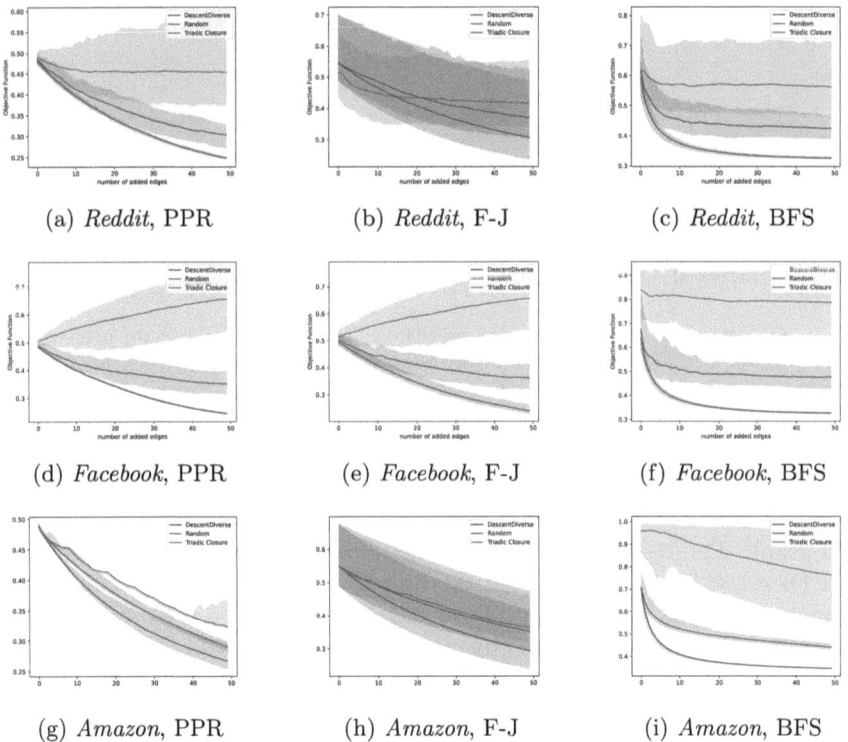

(a) *Reddit*, PPR   (b) *Reddit*, F-J   (c) *Reddit*, BFS

(d) *Facebook*, PPR   (e) *Facebook*, F-J   (f) *Facebook*, BFS

(g) *Amazon*, PPR   (h) *Amazon*, F-J   (i) *Amazon*, BFS

Fig. 5. Objective vs. number of edge additions, inferred communities, uniform target partition.

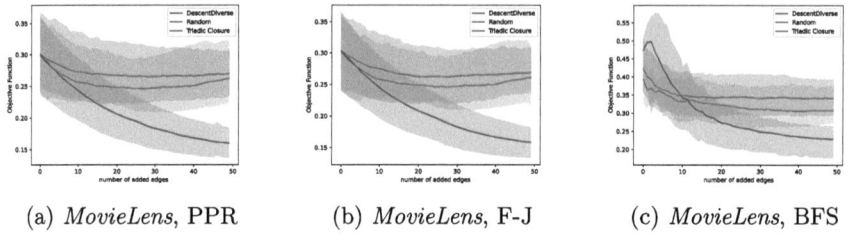

(a) *MovieLens*, PPR   (b) *MovieLens*, F-J   (c) *MovieLens*, BFS

Fig. 6. Objective vs. number of edge additions, *MovieLens*, non-discrete communities.

Non-discrete Communities. Our formulation also allows for optimizing in cases where partition are non-discrete or overlapping. As detailed above, we adapted the *Movielens* dataset so that each movie is defined by a distribution over its movie genres, and optimize for uniform or preferential exposure. The results are shown in Fig. 6. Again, the DESCENTDIVERSE algorithm is the most effective, and TRIADICCLOSE the least effective, in line with the other results.

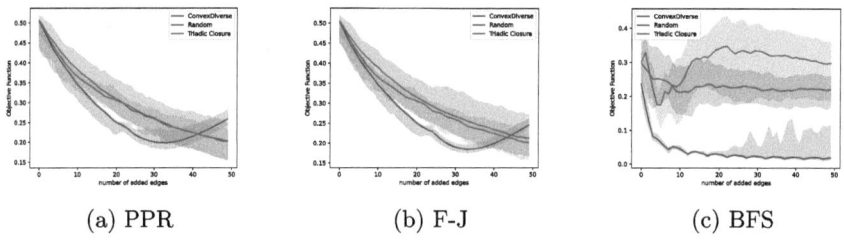

(a) PPR   (b) F-J   (c) BFS

Fig. 7. Objective vs. number of edge additions, *MovieLens User-item*, user-user recommendation, discrete communities, uniform exposure objective.

Recommendation on the User-Item Graph. Figures 7 and 8 present the results for the user-user and user-item diverse recommendation objectives on the user-item graph. The results are similar to those on the original user-user network, with DESCENTDIVERSE achieving the best diversity measure. Notably, recommendations on the user-item graph converge faster than on the user-user graph. For PPR and F-J, the minimal objective is reached after approximately 30 recommendations, while for BFS, it is reached after only 10 recommendations. This speed-up in convergence is likely due to the additional paths in the graph available for optimizing exposure, such as user-item and item-item links. Note that this scenario cannot be directly compared to state-of-the-art baselines, as they are not designed for user-item graphs.

Running Time. The running time of the algorithms, in seconds per iteration, is shown in Table 2. Our algorithms (DESCENTDIVERSE, PARTITIONBOOSTING-

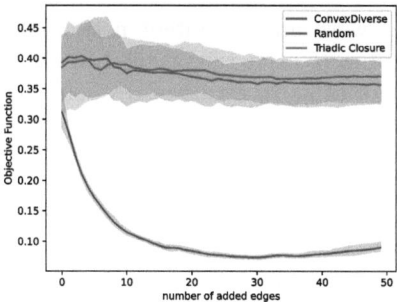

Fig. 8. Objective vs. number of edge additions, *MovieLens User-item*, BFS exposure, user-item recommendation, discrete communities, uniform exposure objective.

Table 2. Running time per iteration in seconds. For *Movielens User-Item*, (u-u) represents user-to-user recommendation, while (u-i) represents user-to-item recommendation.

Dataset	(density)	BFS	PPR	F-J	Greedy	Part. Boost	S.S. Diam. Min.	Global Diam. Min.	Pol. Min.
Polbooks	0.080	0.141	0.217	0.239	2.11	2.37	0.202	0.119	4.29
Football	0.092	0.155	0.247	0.266	2.23	2.03	0.176	0.033	4.99
Movielens	0.186	43.7	18.5	20.5	120	110	10.5	11.7	3064
Reddit	$2.04e^{-4}$	182	136	157	-	-	-	-	-
Facebook	$4.06e^{-4}$	656	513	599	-	-	-	-	-
Amazon	$1.65e^{-5}$	830	2305	2441	-	-	-	-	-
Movielens User-item	$6.09e^{-2}$	(u-u) 3.62	(u-u) 12.9	(u-u) 13.3	-	-	-	-	-
		(u-i) 4,17	-	-	-	-	-	-	-

Diverse, GreedyDiverse) have similar running times. The SpGreedy algorithm is the slowest, but note that we have implemented the exact algorithm in [43], SimpleGreedy, not the fastest, approximate, FastGreedy. The diameter reduction algorithms are fast and provide a reasonable performance/efficiency trade-off, but only for small graphs. For graphs of a few thousand nodes on, the baselines timed-out, but our optimized algorithm is still capable of running on million-node graphs. One interesting result is the peak observed around 10^5 nodes: this is because the dataset used there is a denser graph, so BiCGSTAB requires more iterations to converge.

7   Conclusions and Future Work

We presented in this article a formulation of diversity optimization when keeping into account a node's exposure to information, expressed as a distribution over partitions of opinions in the graph. Our formulation allows to express the problem of local edge additions as an optimization problem, over objectives derived from entropy-based measures. We discuss formalisms for social networks (user-user graphs) but also extend the methods and formalism to hybrid graphs composed of a social network and a content network that users work on. We presented several algorithms to solve the problem, and showed that they outperform state-of-the-art algorithms in terms of diversity measures, and that they are scalable to large graphs.

The fact that our formulation of the objective function is non-monotone opens up the interesting problem of finding the optimal number of edges to add to maximize diversity. Moreover, one should account for the fact that some nodes might not belong to any opinion partition; it is not immediately apparent what is the best semantics to take into account such nodes. Another interesting avenue for research is when the partitions are not known in advance, and hence the diversity measure is a result of a user's feedback in an online algorithm. Finally, while the focus of this paper was not on finding the most efficient algorithm for optimizing diversity of exposure, it is important to study algorithms that exploit massively-parallel matrix computation to achieve truly large-scale diversity optimization.

A   Proofs for Section 5 (Content-Based Extension: User-Item Graph)

Vectorized Transpose Matrix. In the following we will use the Vectorized Transpose Matrix TVEC as defined in [4]. For any vectorized matrix $X \in \mathcal{M}_{m,n}$, the TVEC is the $nm \times nm$ permutation matrix:

$$X^{\top} = \text{TVEC } X. \tag{18}$$

General note Using the product rule and the following formulas for matrix derivatives [4], for $X \in \mathcal{M}_{m,n}$:

$$\frac{\partial(X^T C X)}{\partial X} = \left(I_m \otimes X^{\top} C\right) \partial X + \left(X^{\top} C^{\top} \otimes I_m\right) \partial X^{\top} \tag{19}$$

$$= \left(I_m \otimes X^{\top} C + TVEC[m,m](I_m \otimes X^{\top} C^{\top})\right) \partial X$$

$$\frac{\partial(fg)}{\partial X} = (I \otimes f)\frac{\partial g}{\partial X} + \left(\left(g^{\top} \otimes I\right)\frac{\partial f}{\partial X}\right) \partial X \tag{20}$$

Using $X^{\top} = T$ and adapting M for each exposure function, we then have:

$$\frac{\partial \mathbf{u} T M T^{\top}}{\partial T} = (Id \otimes \mathbf{u})\frac{\partial T M T^{\top}}{\partial T} + \left(T M T^{\top}\right)^{\top} \otimes Id\frac{\partial \mathbf{u}}{\partial T}$$

$$= (Id \otimes \mathbf{u})\left(Id \otimes TM + TVEC[n,n]\left(Id \otimes T^{\top} M^{\top}\right)\right)$$

Proposition 5 (PPR Exposure). *Given the block matrix formulation of a graph* $G = (U \times C)$ *such that* C *has an invertible Schur complement* S *in* G, *then the PPR exposure is given by:*

$$\mathcal{E}_G^{PPR} = (1 - \alpha)\begin{bmatrix} S^{-1} \\ -C^{-1}T^{\top}S^{-1} \end{bmatrix} \tag{21}$$

with $S = \left(I - \alpha U - \alpha^2 T \left(I - \alpha C\right)^{-1} T^{\top}\right)$

Proof.

$$\mathcal{E}_G^{PPR} = W_G^{PPR} \tag{22}$$

$$= (1 - \alpha)\left(I - \alpha G\right)^{-1}$$

$$(I - \alpha G) = \begin{bmatrix} I - \alpha U & \alpha T \\ \alpha T^\top & I - \alpha C \end{bmatrix},$$

$$(I - \alpha G)^{-1} = \begin{bmatrix} S^{-1} & -\alpha S^{-1} T C^{-1} \\ -\alpha C^{-1} T^\top S^{-1} & C^{-1} + \alpha^2 C^{-1} T^\top S^{-1} T C^{-1} \end{bmatrix},$$

$$\mathcal{E}_G^{PPR} \approx (1 - \alpha)\begin{bmatrix} S^{-1} \\ -C^{-1} T^\top S^{-1} \end{bmatrix}$$

$$\text{with } S = \left(I - \alpha U - \alpha^2 T\left(I - \alpha C\right)^{-1} T^\top\right)$$

Proposition 6 (F-J Exposure). *Given the block matrix formulation of a graph $G = (U \times C)$ such that C has an invertible Schur complement S in G, then the F-J exposure is given by:*

$$\mathcal{E}_G^{FJ} = (1 - \alpha)\begin{bmatrix} S^{-1} \\ -C^{-\top} T^\top S^{-1} \end{bmatrix} \tag{23}$$

with $S = \left(I - \alpha U^\top - \alpha^2 T\left(I - \alpha C^\top\right)^{-1} T^\top\right)$

Proof.

$$\mathcal{E}_G^{FJ} = W_G^{FJ} \tag{24}$$

$$= (1 - \alpha)\left(I - \alpha G^\top\right)^{-1}$$

$$(I - \alpha G^\top) = \begin{bmatrix} I - \alpha U^\top & \alpha T \\ \alpha T^\top & I - \alpha C^\top \end{bmatrix},$$

$$(I - \alpha G^\top)^{-1} = \begin{bmatrix} S^{-1} & -\alpha S^{-1} T C^{-\top} \\ -\alpha C^{-\top} T^\top S^{-1} & C^{-\top} + \alpha^2 C^{-\top} T^\top S^{-1} T C^{-\top} \end{bmatrix},$$

$$\mathcal{E}_G^{PPR} \approx (1 - \alpha)\begin{bmatrix} S^{-1} \\ -C^{-\top} T^\top S^{-1} \end{bmatrix}$$

$$\text{with } S = \left(I - \alpha U^\top - \alpha^2 T\left(I - \alpha C^\top\right)^{-1} T^\top\right)$$

Proposition 7 (BFS Exposure). *For any block matrix formulation of a graph $G = (U \times C)$, the BFS (limited to a depth of 3) exposure is given by:*

$$\mathcal{E}_G^{BFS} = \begin{bmatrix} U^3 + U^2 + U + UTT^T + TT^T U + TCT^T + TT^T \\ T^T U^2 + T^T U + T^T TT^T + CT^T U + CT^T + C^2 T^T + T^T \end{bmatrix} \tag{25}$$

$$\mathcal{E}_G^{BFS} = W_G^{BFS} \tag{23}$$

$$= G^3 + G^2 + G$$

$$= \begin{bmatrix} U^3 + U^2 + U + UTT^T + TT^TU + TCT^T + TT^T \\ T^TU^2 + T^TU + T^TTT^T + CT^TU + CT^T + C^2T^T + T^T \end{bmatrix}$$

$$\begin{matrix} U^2T + UT + UTC + TC + TT^TT + TC^2 + T \\ T^TUT + T^TT + T^TTC + CT^TT + C^3 + C^2 + C \end{matrix} \tag{24}$$

$$\mathcal{E}_G^{BFS} \approx \begin{bmatrix} U^3 + U^2 + U + UTT^T + TT^TU + TCT^T + TT^T \\ T^TU^2 + T^TU + T^TTT^T + CT^TU + CT^T + C^2T^T + T^T \end{bmatrix}$$

Proposition 8 (PPR Exposure Derivative). *For the PPR exposure function, the derivatives corresponding to the two problems are given by:*

$$\frac{\partial \mathcal{E}_G^{PPR}}{\partial U} = -\alpha(1-\alpha) \begin{bmatrix} \left(S_{u,u}^{-\top}S^{-1}\right) \\ \left(C^{-1}T^{\top}\right)\left(S_{u,u}^{-\top}S^{-1}\right) \end{bmatrix} \tag{26}$$

$$\frac{\partial \mathcal{E}_G^{PPR}}{\partial T} = (1-\alpha) \begin{bmatrix} -\alpha^2 \left(S_{[u,u]}^{-\top}S^{-1}\right)\left(T(I-\alpha C)^{-1} + (T(I-\alpha C)^{-\top})\right) \\ -\alpha^2 \left(\left(C^{-1}T^{\top}\right)\left(S_{[u,u]}^{-\top}S^{-1}\right)\left(T(I-\alpha C)^{-1} + (T(I-\alpha C)^{-\top})\right)\right) + S_{[u,u]}^{-\top}C^{-1} \end{bmatrix} \tag{27}$$

Proof.

$$\mathcal{E}_G^{PPR} = (1-\alpha) \begin{bmatrix} S^{-1} \\ -C^{-1}T^{\top}S^{-1} \end{bmatrix} \tag{28}$$

with $S = \left(I - \alpha U - \alpha^2 T\left(I - \alpha C\right)^{-1}T^{\top}\right)$

$$d\mathcal{E}_G^{PPR} = (1-\alpha) \begin{bmatrix} \left(S^{-\top} \otimes S^{-1}\right)dS \\ \left(I \otimes C^{-1}T^{\top}\right)\left(S^{-\top} \otimes S^{-1}\right)\,d\,S + \left(S^{-\top} \otimes I\right)\,d\left(C^{-1}T^{\top}\right) \end{bmatrix} \tag{29}$$

$$\frac{\partial \mathcal{E}_G^{PPR}}{\partial U} = -\alpha(1-\alpha) \begin{bmatrix} \left(S^{-\top} \otimes S^{-1}\right) \\ \left(I \otimes C^{-1}T^{\top}\right)\left(S^{-\top} \otimes S^{-1}\right) \end{bmatrix} \tag{30}$$

$$\frac{\partial \mathcal{E}_G^{PPR}}{\partial U}[u] = -\alpha(1-\alpha) \begin{bmatrix} \left(S_{u,u}^{-\top}S^{-1}\right) \\ \left(C^{-1}T^{\top}\right)\left(S_{u,u}^{-\top}S^{-1}\right) \end{bmatrix}$$

$$\frac{\partial \mathcal{E}_G^{PPR}}{\partial T} = (1-\alpha) \begin{bmatrix} -\alpha^2 \left(S^{-\top} \otimes S^{-1}\right)X \\ -\alpha^2 \left(\left(I \otimes C^{-1}T^{\top}\right)\left(S^{-\top} \otimes S^{-1}\right)X\right) + \left(S^{-\top} \otimes I\right)\text{TVEC}(I \otimes C^{-1}) \end{bmatrix} \tag{31}$$

where $X = I \otimes T(I - \alpha C)^{-1} + \text{TVEC}(I \otimes T(I - \alpha C)^{-\top}$.

$$\frac{\partial \mathcal{E}_G^{PPR}}{\partial T}[u] = (1-\alpha) \left[\begin{array}{c} -\alpha^2 \left(S_{[u,u]}^{-\top} S^{-1} \right) \left(T(I - \alpha C)^{-1} + (T(I - \alpha C)^{-\top}) \right) \\ -\alpha^2 \left((C^{-1}T^\top) \left(S_{[u,u]}^{-\top} S^{-1} \right) \left(T(I - \alpha C)^{-1} + (T(I - \alpha C)^{-\top}) \right) \right) + S_{[u,u]}^{-\top} C^{-1} \end{array} \right] \tag{32}$$

Proposition 9 (F-J Exposure Derivative). *For the Friedkin-Johnsen exposure function, the derivatives corresponding to the two problems are given by:*

$$\frac{\partial \mathcal{E}_G^{FJ}}{\partial U}[u] = -\alpha(1-\alpha) \left[\begin{array}{c} S_{u,u}^{-1} S^{-\top} \\ (C^{-\top}T^\top) \left(S_{u,u}^{-1} S^{-\top} \right) \end{array} \right] \tag{33}$$

$$\frac{\partial \mathcal{E}_G^{FJ}}{\partial T}[u] = (1-\alpha) \left[\begin{array}{c} -\alpha^2 \left(S_{[u,u]}^{-\top} S^{-1} \right) \left(T(I - \alpha C^\top)^{-1} + (T(I - \alpha C^\top)^{-\top}) \right) \\ -\alpha^2 \left((C^{-\top}T^\top) \left(S_{[u,u]}^{-\top} S^{-1} \right) \left(T(I - \alpha C^\top)^{-1} + (T(I - \alpha C^\top)^{-\top}) \right) \right) + S_{[u,u]}^{-\top} C^{-\top} \end{array} \right] \tag{34}$$

Proof.

$$\mathrm{d}\mathcal{E}_G^{FJ} = (1-\alpha) \left[\begin{array}{c} \left(S^{-\top} \otimes S^{-1} \right) dS \\ (I \otimes C^{-\top}T^\top) \left(S^{-\top} \otimes S^{-1} \right) \mathrm{d}\, S + \left(S^{-\top} \otimes I \right) \mathrm{d}\, (C^{-\top}T^\top) \end{array} \right] \tag{35}$$

$$\frac{\partial \mathcal{E}_G^{FJ}}{\partial U} = -\alpha(1-\alpha) \left[\begin{array}{c} \text{TVEC} \left(S^{-1} \otimes S^{-\top} \right) \\ (I \otimes C^{-\top}T^\top) \, \text{TVEC} \left(S^{-1} \otimes S^{-\top} \right) \end{array} \right] \tag{36}$$

$$\frac{\partial \mathcal{E}_G^{FJ}}{\partial U}[u] - -\alpha(1 \quad \alpha) \left[\begin{array}{c} S_{u,u}^{-1} S^{-\top} \\ (C^{-\top}T^\top) \left(S_{u,u}^{-1} S^{-\top} \right) \end{array} \right]$$

$$\frac{\partial \mathcal{E}_G^{FJ}}{\partial T} = (1-\alpha) \left[\begin{array}{c} -\alpha^2 \left(S^{-\top} \otimes S^{-1} \right) X \\ -\alpha^2 \left((I \otimes C^{-\top}T^\top) \left(S^{-\top} \otimes S^{-1} \right) X \right) + \left(S^{-\top} \otimes I \right) (I \otimes C^{-\top}) \end{array} \right] \tag{37}$$

where $X = I \otimes T(I - \alpha C^\top)^{-1} + \text{TVEC}(I \otimes T(I - \alpha C^\top)^{-\top}$.

$$\frac{\partial \mathcal{E}_G^{FJ}}{\partial T}[u] = (1-\alpha) \left[\begin{array}{c} -\alpha^2 \left(S_{[u,u]}^{-\top} S^{-1} \right) \left(T(I - \alpha C^\top)^{-1} + (T(I - \alpha C^\top)^{-\top}) \right) \\ -\alpha^2 \left((C^{-\top}T^\top) \left(S_{[u,u]}^{-\top} S^{-1} \right) \left(T(I - \alpha C^\top)^{-1} + (T(I - \alpha C^\top)^{-\top}) \right) \right) + S_{[u,u]}^{-\top} C^{-\top} \end{array} \right] \tag{38}$$

Proposition 10 (BFS Exposure Derivative). *For the BFS exposure function, the derivatives corresponding to the two problems are given by:*

$$\frac{\partial \mathcal{E}_G^{BFS}}{\partial U}[u] = \begin{bmatrix} U_{u,u}^{T^2} I + U_{u,u}^\top U + U^2 + U + U_{u,u}^\top I + TT_{u,u}^\top I + TT^\top \\ T^\top U + U_{i,i}^\top T^\top + T^\top + CT^\top \end{bmatrix} \tag{39}$$

$$\frac{\partial \mathcal{E}_G^{BFS}}{\partial T}[u] = \begin{bmatrix} (UT_{u,u}I) + (T_{u,u}U) + (T_{u,u}U^\top) + (U^\top T_{u,u}I) + TC + TC^\top + T \\ U_{u,u}^{2^\top} I + U_{u,u}^\top I + TT_{u,u}^\top I + T^\top T + U_{u,u}^\top C + C + C^2 \end{bmatrix} \tag{40}$$

Proof.

$$\frac{\partial \mathcal{E}_G^{BFS}}{\partial U} = \begin{bmatrix} U^{T^2} \otimes I + U^\top \otimes U + I \otimes U^2 + I \otimes U + U^\top \otimes I + TT^\top \otimes I + I \otimes TT^\top \\ I \otimes T^\top U + U^\top \otimes T^\top + I \otimes T^\top + I \otimes CT^\top \end{bmatrix} \tag{41}$$

$$\frac{\partial \mathcal{E}_G^{BFS}}{\partial U}[u] = \begin{bmatrix} U_{u,u}^{T^2} I + U_{u,u}^\top U + U^2 + U + U_{u,u}^\top I + TT_{u,u}^\top I + TT^\top \\ T^\top U + U_{i,i}^\top T^\top + T^\top + CT^\top \end{bmatrix}$$

$$\frac{\partial \mathcal{E}_G^{BFS}}{\partial T} = \begin{bmatrix} \text{TVEC}(UT \otimes I) + (T \otimes U) + \text{TVEC}(T \otimes U^\top + (U^\top T \otimes I) + \\ (U^{2^\top} \otimes I) + (U^\top \otimes I) + (TT^\top \otimes I) + \text{TVEC}\,(T^\top \otimes T) + \end{bmatrix} \tag{42}$$

$$\begin{bmatrix} + I \otimes TC + \text{TVEC}(I \otimes TC^\top + I + \text{TVEC}(I \otimes T) \\ + (I \otimes TT^\top) + (U^\top \otimes C) + (I \otimes C) + (I \otimes C^2) \end{bmatrix}$$

$$\frac{\partial \mathcal{E}_G^{BFS}}{\partial T}[u] = \begin{bmatrix} (UT_{u,u}I) + (T_{u,u}U) + (T_{u,u}U^\top) + (U^\top T_{u,u}I) + TC + TC^\top + T \\ U_{u,u}^{2^\top} I + U_{u,u}^\top I + TT_{u,u}^\top I + T^\top T + U_{u,u}^\top C + C + C^2 \end{bmatrix}$$

B   Similarity Between Block Matrix Formulation and Dual Propagation Formulation

We touch briefly on another approach to writing the exposure function, based on a dual propagation approach. For instance, in the case of PPR exposure, we can write the exposure function as a linear combination of the exposures resulting from two propagation processes:

$$\mathcal{E}_G(u) = \theta_1 \mathcal{E}_U(u) + \theta_2 \mathcal{E}_C, \tag{43}$$

where θ_1 and θ_2 are hyperparameters that control the influence of each exposure, and $\mathcal{E}_U$ is the previously defined exposure in the user graph.

The exposure function $\mathcal{E}_U$ is given by:

$$\mathcal{E}_C^{PPR}(u) = W_u^\infty = (1 - \alpha)\mathbf{u}T\,(I - \alpha C)^{-1}\,T^\top,$$

with the resulting overall exposure function:

$$\mathcal{E}_G(u) = \theta_1 \mathcal{E}_U(u) + \theta_2 \mathcal{E}_C(u) \tag{44}$$

$$= \theta_1(1 - \alpha_1)\,(I - \alpha_1 U)^{-1} + \theta_2(1 - \alpha_2)\mathbf{u}T\,(I - \alpha_2 C)^{-1}\,T^\top.$$

We can notice that this exposure has a similar form to that of the Schur complement of C in G using the block matrix formalization $G = (U \times I)$:

$$S = \left(I - \alpha U - \alpha^2 T (I - \alpha C)^{-1} T^\top\right).$$

This shows that we can implement a dual propagation model using block matrix formalization. In this case, the derivatives are the same as those in the previous section.

References

1. Abebe, R., et al.: Opinion dynamics optimization by varying susceptibility to persuasion via non-convex local search. ACM Trans. Knowl. Discov. Data (2022)
2. Bhalla, N., Lechowicz, A., Musco, C.: Local edge dynamics and opinion polarization. In: Proceedings of the Sixteenth ACM International Conference on Web Search and Data Mining, pp. 6–14 (2023)
3. Blondel, V.D., Guillaume, J.L., Lambiotte, R., Lefebvre, E.: Fast unfolding of communities in large networks. J. Stat. Mech. Theory Exper. (2008)
4. Brookes, M.: The matrix reference manual. Imperial College London 3 (2005) (2005)
5. Chen, J., Dong, H., Wang, X., Feng, F., Wang, M., He, X.: Bias and debias in recommender system: a survey and future directions. ACM Trans. Inf. Syst. (2023)
6. Cinelli, M., et al.: The COVID-19 social media infodemic. Sci. Rep. (2020)
7. Cinus, F., Gionis, A., Bonchi, F.: Rebalancing social feed to minimize polarization and disagreement. In: Proceedings of the 32nd ACM International Conference on Information and Knowledge Management (2023)
8. Colin, J., Maniu, S.: Optimizing diverse information exposure in social graphs. In: IEEE International Conference on Big Data (BigData), pp. 519–528 (2024)
9. Demaine, E.D., Zadimoghaddam, M.: Minimizing the diameter of a network using shortcut edges. In: Algorithm Theory - SWAT 2010 (2010)
10. Efstratiou, A., Blackburn, J., Caulfield, T., Stringhini, G., Zannettou, S., De Cristofaro, E.: Non-polar opposites: analyzing the relationship between echo chambers and hostile intergroup interactions on reddit. In: Proceedings of the International AAAI Conference on Web and Social Media (2023)
11. Fabbri, F., Wang, Y., Bonchi, F., Castillo, C., Mathioudakis, M.: Rewiring what-to-watch-next recommendations to reduce radicalization pathways. In: Proceedings of the ACM Web Conference 2022 (2022)
12. Friedkin, N.E., Johnsen, E.: Social Influence and Opinions (1990)
13. Garimella, K., De Francisci Morales, G., Gionis, A., Mathioudakis, M.: Reducing controversy by connecting opposing views. In: Proceedings of the Tenth ACM International Conference on Web Search and Data Mining (2017)
14. Garimella, K., Gionis, A., Parotsidis, N., Tatti, N.: Balancing information exposure in social networks. In: Advances in Neural Information Processing Systems (2017)
15. Ge, Y., et al.: Towards long-term fairness in recommendation. In: Proceedings of the 14th ACM International Conference on Web Search and Data Mining (2021)
16. Ghoshal, A.K., Das, N., Das, S.: Influence of community structure on misinformation containment in online social networks. Knowl.-Based Syst. (2021)
17. Girvan, M., Newman, M.E.J.: Community structure in social and biological networks. In: Proceedings of the National Academy of Sciences (2002)

18. Grover, A., Leskovec, J.: node2vec: scalable feature learning for networks. In: Proceedings of the 22nd ACM SIGKDD International Cconference on Knowledge Discovery and Data Mining (2016)
19. Harper, F.M., Konstan, J.A.: The MovieLens Datasets: history and context. ACM Trans. Interact. Intell. Syst. (2015)
20. Hills, T.T.: The dark side of information proliferation. Perspect. Psychol. Sci. (2019)
21. Interian, R., Moreno, J.R., Ribeiro, C.C.: Polarization reduction by minimum-cardinality edge additions: complexity and integer programming approaches. Int. Trans. Oper. Res. (2021)
22. Isufi, E., Pocchiari, M., Hanjalic, A.: Accuracy-diversity trade-off in recommender systems via graph convolutions. Inf. Process. Manage (2021)
23. Kipf, T.N., Welling, M.: Semi-supervised classification with graph convolutional networks (2016)
24. Kotkov, D., Veijalainen, J., Wang, S.: How does serendipity affect diversity in recommender systems? A serendipity-oriented greedy algorithm. Computing (2020)
25. Kumar, S., Hamilton, W.L., Leskovec, J., Jurafsky, D.: Community interaction and conflict on the web. In: Proceedings of the 2018 World Wide Web Conference (2018)
26. Kunaver, M., Požrl, T.: Diversity in recommender systems – a survey. Knowl.-Based Syst. (2017)
27. Lelkes, Y.: Mass Polarization: Manifestations and Measurements. Public Opinion Quarterly (2016)
28. Levy, R.: Social Media, News Consumption, and Polarization: evidence from a field experiment. Am. Econ. Rev. (2021)
29. Mansoury, M., Abdollahpouri, H., Pechenizkiy, M., Mobasher, B., Burke, R.: FairMatch: a graph-based approach for improving aggregate diversity in recommender systems. In: Proceedings of the 28th ACM Conference on User Modeling, Adaptation and Personalization (2020)
30. Matakos, A., Aslay, C., Galbrun, E., Gionis, A.: Maximizing the diversity of exposure in a social network. IEEE Trans. Knowl. Data Eng. (2022)
31. Musco, C., Musco, C., Tsourakakis, C.E.: Minimizing polarization and disagreement in social networks. In: Proceedings of the 2018 World Wide Web Conference on World Wide Web - WWW '18 (2018)
32. Page, L.: The PageRank citation ranking: bringing order to the web. Technical report. Stanford Digital Library Technologies Project, 1998 (1998)
33. Pitoura, E., Stefanidis, K., Koutrika, G.: Fairness in rankings and recommendations: an overview. The VLDB J. (2022)
34. Prasetya, H.A., Murata, T.: A model of opinion and propagation structure polarization in social media. Comput. Soc. Netw. 7(1), 1–35 (2020). https://doi.org/10.1186/s40649-019-0076-z
35. Ross Arguedas, A., Robertson, C., Fletcher, R., Nielsen, R.: Echo chambers, filter bubbles, and polarisation: a literature review. Reuters Institute for the Study of Journalism (2022)
36. Sasahara, K., Chen, W., Peng, H., Ciampaglia, G.L., Flammini, A., Menczer, F.: Social influence and unfollowing accelerate the emergence of echo chambers. J. Comput. Soc. Sci. (2021)
37. Stewart, A.J., Mosleh, M., Diakonova, M., Arechar, A.A., Rand, D.G., Plotkin, J.B.: Information gerrymandering and undemocratic decisions. Nature (2019)

38. Viswanath, B., Mislove, A., Cha, M., Gummadi, K.P.: On the evolution of user interaction in Facebook. In: Proceedings of the 2nd ACM Workshop on Online Social Networks (2009)
39. Wang, Y., Ma, W., Zhang, M., Liu, Y., Ma, S.: A survey on the fairness of recommender systems. ACM Trans. Inf. Syst. (2023)
40. Yang, J., Leskovec, J.: Defining and evaluating network communities based on ground-truth. In: ACM SIGKDD Workshop on Mining Data Semantics (2012)
41. Yin, K., Fang, X., Chen, B., Sheng, O.R.L.: Diversity preference-aware link recommendation for online social networks. Inf. Syst. Res. (2023)
42. Zhao, Z., et al.: Recommending what video to watch next: a multitask ranking system. In: Proceedings of the 13th ACM Conference on Recommender Systems (2019)
43. Zhu, L., Bao, Q., Zhang, Z.: Minimizing polarization and disagreement in social networks via link recommendation. In: Advances in Neural Information Processing Systems (2021)
44. Ziarani, R.J., Ravanmehr, R.: Serendipity in recommender systems: a systematic literature review. J. Comput. Sci. Technol. **36**(2), 375–396 (2021). https://doi.org/10.1007/s11390-020-0135-9
45. Zinkevich, M.: Online convex programming and generalized infinitesimal gradient ascent. In: International Conference on Machine Learning, pp. 928–935 (2003)

SISIS: Sequence Indexing for SImilarity Search

Sara Jarrad[✉] [ID], Hubert Naacke [ID], and Stéphane Gançarski [ID]

LIP6, Sorbonne University, Paris, France
{sara.jarrad,hubert.naacke,stephane.gancarski}@lip6.fr

Abstract. Sequence similarity computation aims to quantify how similar two sequences are by assigning a similarity score. In this study, we focus on the problem of identifying sequences that are similar to a given query sequence provided by a user. Measuring this similarity is important in several applications, such as in mobility field, where analyzing users' travel patterns can help improve traffic management. Other examples include the classification of similar users and the recommendation of points/sequences. In those examples, the size of the sequences is rarely over ten points.

Existing methods predominantly rely on applying a similarity function to each candidate sequence to identify those that are sufficiently similar. However, this approach becomes computationally expensive when dealing with large-scale datasets. To mitigate this challenge, we propose *SISIS*, an efficient method that uses sequence indexing to quickly retrieve similar sequences that share common points with the query sequence in the same order. Furthermore, to account for scenarios where points in sequences may not exactly match but are contextually similar, we introduce *SISIS**, a variant of SISIS that incorporates point embeddings. This extension allows for more comprehensive retrieval of similar sequences by considering semantics similarities between points, beyond mere exact matches.

Extensive experimental evaluations show that the proposed approach significantly outperforms a baseline method based on the well-known Longest Common SubSequence (LCSS) measure for all reasonable query sizes up to 17, *i.e.* sequences containing 17 points, yielding substantial performance improvements across various real-world datasets.

Keywords: Sequence · Similarity search · Sequence indexing

1 Introduction and Motivation

The amount of information published on the web has been steadily increasing over the years. Among this vast amount of data, geolocalized information shared by users on social networking platforms—through photos, tags, comments, and other interactions—plays a crucial role in describing mobility behavior. This data is particularly valuable as it enables the reconstruction of user trajectories, which are defined as sequences of points (locations) visited by a user in chronological order.

ⓒ The Author(s), under exclusive license to Springer-Verlag GmbH, DE, part of Springer Nature 2026
A. Hameurlain et al. (Eds.): *Transactions on Large-Scale Data- and Knowledge-Centered Systems LIX*, LNCS 16240, pp. 32–60, 2026.
https://doi.org/10.1007/978-3-662-72449-1_2

Measuring sequence similarity is a core task in many applications, as it quantifies how alike two sequences are. This capability is particularly useful for retrieving sequences that are similar to a given query sequence q, typically of reasonable length— containing fewer than five or six points. In the context of mobility, it can support tasks such as user classification or the recommendation of locations/sequences, among others.

Our goal is to develop a similarity measure that is both accurate and efficient. However, this task is inherently challenging, and numerous approaches have been proposed to tackle it, each with its own strengths and limitations. Choosing the right similarity measure ultimately depends on the specific needs and constraints of the application.

Many research studies have focused on quantifying the similarity between two sequences using established measures such Edit Distance on Real Sequence (EDR) [7], and Edit Distance with Real Penalty (ERP) [6]. These measures are designed on edit operations to handle noise-related issues, gaps, or local misalignments among sequences. However, they have a major limitation which is their high computational cost, especially when they are used on large datasets.

Dynamic Time Warping (DTW) [2,28] constitutes another extensively used measure. It aligns time series by minimizing the cumulative distance between their points through a flexible time warping path. While DTW is computationally outperforming exhaustive edit-based measures, it relies fundamentally on pointwise distance computations, most commonly the Euclidean distance. However, it is not applicable to our case, as we do not consider euclidean distances between points. Our focus is solely on the sequential nature of the data, while disregarding its spatio-temporal dimension. This choice allows us to generalize an appropriate approach to various types of sequences beyond mobility sequences. Longest Common SubSequence (LCSS) [12,13,20], is a measure that can be applied to sequential only data. It is the length of the longest common subsequence of points within two sequences, where two compared points are considered to match only if they are identical.

Fig. 1. Use case of LCSS on mobility sequences

Figure 1 illustrates a simplified map of Paris city, highlighting its most renowned locations. An illustrative example of a user sequence is depicted by the sequence of points [2, **6, 8, 3**, 4, 7], represented by a grey dotted path. This sequence corresponds to a user starting their visit at the Orsay Museum, identified by ID 2, followed by the Eiffel Tower (ID 6), then the Louvre Museum (ID 8), and so forth. The objective is to retrieve sequences similar to a given query sequence [**6, 8, 3**], shown as a red path. One potential solution to address this problem is the use of the Longest Common Subsequence (LCSS) measure.

LCSS-based similarity is widely regarded as one of the most effective measures due to its robustness against various transformations that may be applied to sequences, such as re-sampling (adding points to sequences) and handling data noise (introducing outliers in the sequences) [16, 22, 26].

Despite these advantages, computing the LCSS-based similarity can be computationally expensive when applied to large datasets of sequences. In a naive approach (baseline) for identifying sequences similar to a given query sequence q, the LCSS between q and each candidate sequence is calculated exhaustively. This inefficiency in processing time represents a significant challenge, and is one of the primary issues we address in this work.

To address it, we propose *SISIS* (Sequence Indexing for SImilarity Search). SISIS is an efficient method that employs a sequence indexing approach to accelerate the search for similar sequences. This approach achieves the same results as the LCSS-based baseline method (see Algorithm 2), while significantly improving search speed. Moreover, SISIS allows users to specify a desired similarity threshold, defined as the number of points in common with the query, in the same sequential order.

A second issue we tackle is that exact matching based on point identification can be overly restrictive, potentially resulting in very few or no matching sequences. In many practical scenarios, users may not require exact point matches but instead seek sequences whose points are contextually similar to those in the query.

In order to meet this challenge, we introduce *SISIS**, a variant of SISIS that leverages point embeddings generated by the Word2Vec (W2V) language model [17], to enhance the retrieval of similar sequences, even when their points do not exactly match but are contextually similar. SISIS* enables a broader set of results compared to the exact matching performed by the original SISIS method.

The remainder of the paper is structured as follows. Sections 2 and 3 provide background on LCSS and review related work. Section 4 describes the exact sequence indexing approach, while Sect. 5 explores an alternative indexing method using point embeddings. Section 6 reports the experimental results and performance comparisons. Finally, Sect. 7 concludes the paper.

2   Background

The algorithm described in Algorithm 1 has been detailed in [12]. It computes the size of the longest common subsequence between two sequences. This size is later used to compute the similarity between sequences. We extend the original version by adding the point matching function as a parameter.

Algorithm 1 LCSS size

Require: q: query sequence, t: candidate sequence, $match$: point matching function

1: **function** LCSS($q, t, match$)
2:     $\forall i \in [1, |q|], j \in [1, |t|], M[i][j] \leftarrow 0$ // Initialize the similarity matrix
3:     **for** i in $[1, |q|]$, j in $[1, |t|]$ **do**
4:         **if** $match(q[i-1], t[j-1])$ **then**
5:             $M[i][j] \leftarrow M[i-1][j-1] + 1$
6:         **else**
7:             $M[i][j] \leftarrow \max(M[i-1][j], M[i][j-1])$
8:         **end if**
9:     **end for**
10:    **return** $M[|q|][|t|]$ // length of the LCSS
11: **end function**

$LCSS(q, t)$ denotes $LCSS(q, t, =)$ *i.e.* the default $match$ is the equality function and two points being compared only match if they are the same. *i.e.* if they share the same unique identifier in the dataset.

Example 1. Consider query $q = [A, \mathbf{D}, B, \mathbf{E}, \mathbf{C}]$, and a sequence $t = [F, \mathbf{D}, G, \mathbf{E}, H, \mathbf{C}, A]$, then the longest common sub-sequence is $[\mathbf{D}, \mathbf{E}, \mathbf{C}]$ and $LCSS(q, t) = 3$.

2.1 Problem Statement

Let $T = \{t_1, \cdots, t_n\}$ be a set of sequences. Let q be a query sequence, and $t \in T$ a sequence from the dataset. We denote by $LCSS(q, t)$ the length of the longest common subsequence between q and t computed by Algorithm 1.

We consider a sequence t to be similar to q for a given similarity threshold $S \in [0, 1]$ if its longest common subsequence contains at least $cs = \lceil |q| \times S \rceil$ points. In other words, the ratio of the sizes of q and LCSS(q,t) must be greater than S. We denote $q \approx_S t$ such a similarity property defined by:

$$q \approx_S t \equiv \frac{LCSS(q, t)}{|q|} \geq S$$

We consider a user submitting a query composed of an input sequence q, and a similarity threshold S. We aim to retrieve the set $T'(q)$ of all sequences in T that are similar enough to q, that is $T'(q) = \{t \in T | q \approx_S t\}$.

Example 2. Let us consider a query sequence $q = [A, B, C, D, E]$ and two candidate sequences: $t_1 = [K, \mathbf{A}, F, \mathbf{D}]$ and $t_2 = [M, O, \mathbf{A}, \mathbf{B}, F, \mathbf{C}, P, \mathbf{E}]$. Suppose the user specifies a similarity threshold $S = 0.6$, corresponding to $cs = 3$. As per the definitions provided above, sequence t_2 is considered similar to q under the given threshold S, as $\frac{\text{LCSS}(q,t_2)}{|q|} = \frac{4}{5} \geq S$. In contrast, sequence t_1 does not meet the similarity threshold.

2.2 LCSS-Based Baseline

As mentioned before, to determine the sequences that are similar to a given query q, the LCSS between q and all other sequences is calculated (Algorithm 2). Thus, the LCSS function defined in Algorithm 1 is invoked $|T|$ times.

Algorithm 2 LCSS-based search

S

Require: q: query sequence, S: threshold, $match$: point matching function

```
1: function LCSS_SEARCH(q, S, match)
2:     result ← ∅,   cs ← ⌈|q| × S⌉
3:     for t in T do
4:         if LCSS(q, t, match) ≥ cs then
5:             result ← t
6:         end if
7:     end for
8:     return result
9: end function
```

On lines 4 and 5 of Algorithm 2 we apply the LCSS to all dataset sequences, and select one if it shares at least cs points with q. This process incurs a significant computational overhead, particularly when the dataset is very large, and we are looking to reduce it. We use Algorithm 2 as the main baseline for comparison throughout the paper. Our goal is to develop a new method, called SISIS, that retrieves similar sequences faster than this baseline.

The notations frequently used throughout this paper are summarised in Table 1.

3 Related Work

There are several studies that address the problem of finding similar sequences, especially in the field of mobility. For instance [16] categorizes different similarity measures, discussing their respective strengths and weaknesses with regard to noise sensitivity, time shift (when an element of one sequence is shifted in time to match an element of another sequence) and sequence length. These measures are classified according to whether they are spatial *i.e.* ignoring the temporal aspect, or spatio-temporal, taking into account both the spatial and temporal dimensions of motion. Spatial measures include Euclidean, Hausdroff, and Fréchet distances. The first does not allow to compare sequences of different sizes and is not robust to noise. The second and third are based solely on the geographical shape of the sequences being compared, and are not robust to data noise or time shifts. There are other similarity measures, such as DTW, EDR and ERP, which are among the most well-known, but neither are they robust to noise.

In [16], LCSS is considered one of the best measures because it is robust to almost all transformations compared to the other measures.

Table 1. Notations used for similarity search

Notation	Description
q	A query sequence
cs	Combination size
T	Set of sequences
$T'(q)$	Set of all sequences that are similar to the query q
$q \approx_S t$	a sequence t is similar to q for a given threshold S
$pos(p_i, t)$	The position of the point p_i in a sequence t
$C_{|q|,cs}$	the set of $C_{cs}^{|q|}$ combinations of size cs in the query q
$\overline{p_i}$	An index that associates each element p_i with the sequences that contain it
$\overline{p_i p_j}$	An index that associates each pair of elements $p_i p_j$ with the sequences that contain it
$sim_\epsilon(p_i, p_j)$	p_i and p_j are ϵ-similar
$\overline{\overline{p_i}}$	An index that associates each point p_i with the set of sequences that pass through a point p_j ϵ-similar to p_i

[22, 26] compare the same measures, but categorizes them differently depending on whether they are spatio-temporal or sequence-only measures, and whether they are discrete or continuous. To evaluate the robustness of each measure, transformations such as adding/removing sample points, altering the sampling rate, injecting noise (outliers), and shifting sequences were applied to the data. The authors then assessed how well each measure handled these modifications. They concluded that DTW, LCSS, and ERP were the most effective across various transformation types. Table 2 summarizes these findings. Each checkmark (✓) in the table indicates that the corresponding measure performs well under the given transformation scenario.

Table 2. Comparative study of similarity measures from [22, 26]

Distance Measure	Transformation					
	Adding sample points	Deleting sample points	Different sampling rate	Spatial stretch and suppress	Adding noise	Mixed transformations
ED						✓
DTW	✓	✓	✓	✓		✓
LCSS	✓	✓	✓	✓	✓	✓
EDR					✓	
ERP	✓	✓	✓	✓	✓	

The DTW and ERP measure demonstrates the ability to handle at least four types of transformations, with LCSS standing out as the most effective in managing all transformations.

A comparative study has also been proposed by [24], which examines the four similarity measures LCSS, Fréchet, DTW and EDR. This paper highlights the differences between these four measures using real data and also shows some popular applications of these measures.

Apart from the mobility field, LCSS is used in biology for DNA alignment. Other methods which have the same purpose as LCSS are, local alignment [9,21], global alignment [18,21], multiple alignment [19], and BLAST [1]. Local alignment identifies common elements between two sequences, even if they do not overlap over their entire length. As for the global alignment, the objective is to maximize the overall similarity through the sequence alignment by taking into consideration the gaps between elements, while preserving the order of these elements. Multiple alignment aligns multiple sequences simultaneously to find common motifs among them all. Finally, Blast is used to find regions of similarity between biological sequences based on local alignment. All of these measures are effective, but they are all complex on a large scale.

In summary, there are a variety of methods, using different approaches to solve sequence similarity search problem, and the works cited explore the differences between them. All methods have advantages and disadvantages. However, according to the experimental validation used in [16,22,26], LCSS remains one of the most robust to various mentioned transformations. Thus we base our work on this measure.

The issue of those cited works is that they do not address the cost problem. In fact, these methods are effective when applied to a small dataset. Once scaled up, they become very costly and time consuming.

[5] uses efficient algorithms for finding a longest common increasing subsequence among m sequences. However, the authors assume that the sequences of points they use are increasing, *i.e.* the points are ordered by increasing values. In our case, we represent a point by its id, so, if the points were ordered by their values, the order of the point visits would be lost. Therefore, the solution is not applicable to our problem.

[11] uses a dual-match method for matching sub-sequences. Their aim is to find all the subsequences (from a very large data sequence) that are similar to a given input query. The authors of [11] propose a complex solution, by slicing the data sequence and the query into windows, then representing it in the form of multi-dimensional vectors, then applying the LCSS using a parameter to match intervals in the time dimension. They apply their method to time series data: their index is based on the data value and the elapsed time between two consecutive points, which is not the case for us.

Some other recent works such as [4,10] have proposed solutions to approximate LCSS in linear time (instead of quadratic time as for Algorithm 1) with respect to the size of sequences. This makes the computation time of LCSS faster. However, the problem with these approaches is that they do not give the exact results provided by LCSS, but rather an approximation using probability to make a tradeoff between accuracy and efficiency.

Whereas these works compute LCSS whatever the size of the longest common subsequence, we study a different problem where the minimal size of the sub-sequence to retrieve is specified by the user.

4   Exact Sequence Indexing

This section introduces the main contribution of this work: an indexing method designed for the efficient retrieval of similar sequences.

Let $T = \{t_1, \ldots, t_n\}$ represent a set of sequences, and let $T'(t, S)$ denote the subset of sequences in T that are sufficiently similar to a given sequence t, based on a specified similarity threshold S, as formally defined in Sect. 2.1. The proposed method employs sequence indexing to efficiently compute $T'(t, S)$.

The key advantage of this indexing approach, compared to the baseline method outlined in Algorithm 2, which performs pairwise comparisons of each sequence in T with t, is its ability to reduce the number of candidate sequences. By narrowing down the search space, the indexing method is expected to significantly improve retrieval speed for queries of a reasonable size.

4.1 Definition of Sequence Index

Definition 1 (Single point index). *The sequence index $1P : p_i \mapsto \bar{p}_i$ associates each point p_i with the set $\bar{p}_i$, which consists of all sequences that pass through p_i. This set is formally defined as:*

$$\bar{p}_i = \{t \in T \mid p_i \in t\}$$

Definition 2 (Point pair index). *The sequence index $2P : (p_i, p_j) \mapsto \overline{p_i p_j}$ associates a pair of points (p_i, p_j) with the set $\overline{p_i p_j}$ of sequences that pass through p_i then through p_j (i.e. p_i precedes p_j) and defined by:*

$$\overline{p_i p_j} = \{t \in T \mid p_i, p_j \in t \text{ and } pos(p_i, t) < pos(p_j, t)\}$$

where $pos(p, t)$ is the position of the point p in sequence t.

The index structure is implemented as a dictionary, where each point or pair of points is mapped to the set of corresponding sequences.

4.2 Index-Based Similarity Search

We first retrieve the $\bar{p}_i$ sequences for each point in the query q. Since the S threshold indicates that the user expects to obtain sequences that share at least $cs = \lceil |q| \times S \rceil$ points with q, we generate all the sub-sequences of q of size cs. For each sub-sequence, we obtain candidate results by computing the intersection of the sets $\bar{p}_i$ corresponding to the points in that sub-sequence. Finally, we verify whether the points in the sub-sequence appear in the right order in the candidate sequence, otherwise the candidate is filtered out.

We provide a detailed explanation of Algorithm 3 which retrieves similar sequences. The algorithm takes as input a query q, a similarity threshold S, the created index I_1, and the point matching function $match$.

Algorithm 3 Find similar sequences

Require: q: query sequence, S: threshold, I_1: single point index, $match$: point matching function

 1: **function** SIMILAR_SEQUENCES($q, S, I_1, match$)
 2:     $result \leftarrow \emptyset, \quad cs \leftarrow \lceil |q| \times S \rceil$
 3:     $C_{|q|,cs} \leftarrow$ the set of $C_{cs}^{|q|}$ combinations of size cs in q
 4:     **for** $combi \in C_{|q|,cs}$ **do**
 5:         $candidates \leftarrow \bigcap_{p_i \in combi} \bar{p}_i$ // use of I_1 index
 6:         **for** $c \in candidates$ **do**
 7:             **if** $c \notin result$ and $same\_order(c, combi, match)$ **then**
 8:                 $result \leftarrow result \cup \{c\}$
 9:             **end if**
10:         **end for**
11:     **end for**
12:     **return** $result$
13: **end function**

On line 2, the minimum number of points, cs, is computed from the user's threshold S. On line 3, we generate the set of $C_{cs}^{|q|}$ combinations of size cs from the query sequence q. For each combination, denoted as $combi$, we compute the set of candidate sequences that contain all the points in $combi$ by intersecting the sets $\bar{p}_i$ for which $p_i \in combi$. Finally on lines 7–9, we verify that the order of the points in each candidate sequence matches the order in $combi$ (as detailed in Algorithm 4).

Algorithm 4 takes as input a candidate sequence c, a combination $combi$, and a point matching function (with point equality by default). The algorithm iterates over the points in c and $combi$, comparing them consecutively: if the points at the current positions i in c and j in $combi$ are equal, the match count is incremented, and the algorithm proceeds to compare the points at positions $i + 1$ and $j + 1$. Otherwise, *i.e.* if the points do not match, the next point of the candidate sequence (at position $i + 1$) is compared with the current point of $combi$ (at position j). After going through all the points in $combi$, the algorithm returns $True$ if all the points of $combi$ were matched, *i.e.* if the number of matches m equals the size of $combi$.

To ensure reproducibility, the code implementing our solution is available at [1]

4.3   Indexing Sequences by Pairs of Points

To enhance the performance (*i.e.* reduce the response time) of similar sequence retrieval, we extend the above indexing method to access sequences containing not just a single point, but pairs of points. Since a point pair index (2P index, see Definition 2) is more selective than a single point index (1P index), we expect to get a performance benefit at the price of a larger index that is longer to build. However, the index creation time is amortized along the index usage period, thus this solution is considered beneficial as

---

[1] https://gitlab.lip6.fr/jarrad/tisis.

Algorithm 4 Check identical points order

Require: c: candidate sequence, $combi$: sub-sequence, $match$: point matching function

1: **function** SAME_ORDER($c, combi, match$)
2:    $i \leftarrow 0, \quad j \leftarrow 0, \quad m \leftarrow 0$
3:    **while** $i < |c|$ **and** $j < |combi|$ **do**
4:       **if** $match(c[i], combi[j])$ **then**
5:          increment j, increment m
6:       **end if**
7:       increment i
8:    **end while**
9:    **return** ($m = |combi|$)
10: **end function**

---

long as it improves sequence retrieval time. Information on the 1P and 2P index sizes, construction times, and memory usage can be found in Sect. 6.4, specifically in Table 5.

Assuming a 2P index, we can access the set $\overline{p_i p_j}$ of sequences that pass through p_i followed by p_j (*i.e.* p_i precedes p_j). We adapt Algorithm 3 to access sequences by the point pair index. Specifically, on line 5, the index is probed for every pair of consecutive points in $combi$. Algorithm 5 is an adaptation of Algorithm 3.

Algorithm 5 Find similar sequences using 2P index

Require: q: query sequence, S: threshold, I_2: Pair of points index

1: **function** SIMILAR_SEQUENCES_2P(q, S, I_2)
2:    $result \leftarrow \emptyset, \quad cs \leftarrow \lceil |q| \times S \rceil$
3:    $C_{|q|,cs} \leftarrow$ the set of $C_{cs}^{|q|}$ combinations of size cs in q
4:    **for** $combi \in C_{|q|,cs}$ **do**
5:       $candidates \leftarrow \displaystyle\bigcap_{p_i, p_j \in combi, \ pos(p_j, combi) = pos(p_i, combi)+1} \overline{p_i p_j}$ // use of I_2 index
6:       $result \leftarrow result \cup candidates$
7:    **end for**
8:    **return** $result$
9: **end function**

4.4 Theoretical Costs

In this section, we analyze the estimated theoretical cost of two approaches:

– SISIS, which refers to the execution of Algorithm 3, and whose cost depends on five key parameters: the query size $|q|$, the combination size cs, the average intersection computation time iat, the average filtering time fat, and the average size of candidate sequences as

– The baseline, which refers to the call to Algorithm 2, and whose estimated cost is determined by the following parameters: the average size of sequences in the dataset t_{as}, the average time to compute similarity between 2 points a_t, and m the total number of sequences to be compared with the query q.

All the mentioned parameters that are used in this analysis are shown in Table 3.

Table 3. Notations used for theoretical cost formulas

Notation	Description
$|q|$	Query size
cs	Combination size
iat	Intersection average computation time
fat	Filtering average computation time
as	Average size of candidate sequences
t_{as}	Average size of sequences in the dataset
a_t	Average time to compute similarity between 2 points
n	Total number of sequences to be compared with q

Theoretical Costs of an Index-Based Similarity Search. The proposed index-based approach follows these steps:

1. **Sequence similarity search:** This part consists of the following sub-steps:
   (a) **Generating point combinations:** We generate the set of $C_{cs}^{|q|}$ combinations of size cs from the query sequence q. (line 3 of Algorithm 3). The cost of generating combinations is negligible.
   (b) **Index access:** For each combination $combi$, we retrieve the sets $\bar{p}_i$ for all $p_i \in combi$ (lines 4–5 of Algorithm 3). Each retrieval is done in constant time, thanks to the pre-constructed index. Since each combination has cs elements and there are $C_{cs}^{|q|}$ such combinations, this results in a cost $AC = cs \cdot C_{cs}^{|q|} \cdot probe$ for index accesses, where $probe$ is the constant index access time.
   (c) **Computing intersections:** For each combination, denoted as $combi$ containing cs elements, we compute the set of candidate sequences that contain all the points of $combi$, by intersecting the sets $\bar{p}_i$ for which $p_i \in combi$, retrieved in the previous sub-step (line 4–5 of Algorithm 3). In total, we calculate $cs - 1$ intersections. The cost of an intersection between two sets of size $s1$ and $s2$ is $\mathcal{O}(s1 * s2)$. However, in our case we can't estimate the exact sizes of $\bar{p}_i$. We choose to measure the run time of intersections and use the average cost iat as an approximation of the intersection cost.

The cost of computing intersections is therefore: $iat \times (cs - 1)$.
The total cost of the sequence similarity search is:

$$SC = (cs - 1) * C_{cs}^{|q|} * iat$$

2. **Sequence filtering:** The final step involves filtering sequences, as described in lines 7–9 of Algorithm 3 (for more details, see Algorithm 4). Given a set of candidate sequences *candidates*, we check whether the order of points in each sequence matches the order in *combi*. The size of *candidates* varies depending on the combination, so we compute the average size (*as*). For each sequence in *candidates*, we measure the filter time (*ft*). We then calculate the average (*fat*) from all the calculated durations. Thus, the filtering cost would be:

$$FC = fat * as * C_{cs}^{|q|}$$

As a result, the estimated total cost of SISIS is:

$$
\begin{aligned}
TC(SISIS) &= AC + SC + FC \\
&= (cs * C_{cs}^{|q|} * probe) + ((cs - 1) * C_{cs}^{|q|} * iat) + (fat * as * C_{cs}^{|q|}) \\
&= C_{cs}^{|q|} * ((cs * probe) + ((cs - 1) * iat) + (fat * as)) \quad (1)
\end{aligned}
$$

When $C_{cs}^{|q|}$ is maximized, *i.e.* $cs = |q|/2$, we have $C_{\frac{|q|}{2}}^{|q|} \approx \frac{2^{|q|}}{\sqrt{|q|}}$. Then, the SISIS cost is $\mathcal{O}(2^{|q|})$ according to Eq. (1).

Theoretical Costs of the Baseline. We also analyze the estimated theoretical cost of the LCSS-based baseline. When a user gives a query q as input, and wants to retrieve similar sequences, the baseline compares the similarity between q and all other sequences in the dataset, and then keeps the similar ones within a certain threshold set by the user.

The baseline first compares the query q of size $|q|$ with a sequence t of average size t_{as}. Since all points in both sequences are compared, the computational cost of this operation is $|q| \times t_{as} \times a_t$, where a_t represents the average time required to compute the similarity between two points. This operation is performed n times, where n denotes $|T|$ *i.e.* the total number of sequences to be compared with q in the dataset. Consequently, the estimated total cost is given by:

$$TC(Baseline) = |q| * t_{as} * a_t * n \quad (2)$$

According to Eq. (2), the LCSS-based baseline has a theoretical cost of $\mathcal{O}(|q|)$ for a fixed dataset, meaning that the cost increases linearly with the query size. This assumes that the number of sequences to be compared (n) and the average sequence size (t_{as}) remain constant.

While this linear complexity appears favorable in theory, experimental results (see Sect. 6.4) show that SISIS can be more efficient for moderate query sizes, typically when $|q|$ is less than 17 or 18, which is the case for all mobility based applications. This conclusion is supported by applying the cost models from Eq. (1) and Eq. (2), using the parameter values listed in Table 3, derived from empirical measurements.

5   Sequence Indexing Based on Point Embeddings

As discussed in Sect. 1, both the LCSS-based baseline and SISIS are effective in finding similar sequences, but they rely on exact point matching. In some cases, this constraint can be too restrictive, leading to a small or even empty result set, particularly for large queries.

To relax this constraint, and assuming that the only available contextual information about points is the sequences that contain them, we propose a method based on point embeddings generated using Word2Vec (W2V) [17]. The principle of W2V is that words that appear in similar contexts tend to have similar vector representations. The model learns this by predicting surrounding words of a target word (or the reverse), leading to similar vectors for words that are used in similar ways.

In our scenario, we can think of the words as being the points and the sentences as being the sequences containing these points. We argue that two sequences are similar not only when their points match, but also when their points have similar embeddings.

Based on this approach, we introduce SISIS*, an extension of SISIS that leverages embeddings. This variant takes into account a relaxation threshold specified by the user, and compares sequences by exploiting contextual similarity. This method provides more results than SISIS. Those additional sequences do not necessarily pass through the query points but through points that are close to them in the embeddings space, *i.e.* they are *contextually* close.

5.1   Contextual Similarity and Sequence Index

Let p_i and p_j two points, and e'_i and e'_j their respective embedding obtained by W2V [17].

Definition 3 (Point ϵ-similarity). *We say that p_i and p_j are ϵ-similar if the cosine between e_i and e_j is above a similarity threshold $\epsilon \in [0,1]$ given by the user. We define sim_ϵ as:*

$$sim_\epsilon(p_i, p_j) \equiv cosine(e_i, e_j) \geq \epsilon$$

Specifically for $\epsilon = 1$, p_i and p_j are the same point, and thus they are similar.

The similarity between two sequences is defined in Sect. 2.1, but considering a contextual version of LCSS, denoted $LCSS(q, t, sim_\epsilon)$, where two points match if they are ϵ-similar. We now generalize the definition of the index.

Definition 4 (Contextual sequence index). *The contextual sequence index $CSI : p_i \mapsto \bar{\bar{p}}_i$ associates each point p_i with the set $\bar{\bar{p}}_i$ of sequences that pass through a point p_j ϵ-similar to p_i and defined by:*

$$\bar{\bar{p}}_i = \{t \in T | \exists p_j \in t, \ s.t. \ sim_\epsilon(p_i, p_j)\}$$

5.2   Sequence Search Based on Point Embeddings

This section details the method we designed to integrate point embeddings in sequence search.

We trained W2V using all the sequences as input and obtained as output vector representations for each point, called point embeddings. These embeddings are used to create the contextual sequence index $\bar{\bar{p}}_i$ and to identify similar sequences as outlined in Algorithm 3. The adaptation made is in Algorithm 6, provided that line 5 becomes

$$candidates \leftarrow \bigcap_{i \in combi} \bar{\bar{p}}_i$$

Algorithm 6 Find similar sequences

Require: q: query sequence, S: threshold, sim_ϵ: point matching function, CSI: contextual sequence index

1: **function** SIMILAR_SEQUENCES(q, S, CSI, sim_ϵ)
2:     $result \leftarrow \emptyset, \quad cs \leftarrow \lceil |q| \times S \rceil$
3:     $C_{|q|,cs} \leftarrow$ the set of $C_{cs}^{|q|}$ combinations of size cs in q
4:     **for** $combi \in C_{|q|,cs}$ **do**
5:         $candidates \leftarrow \bigcap_{i \in combi} \bar{\bar{p}}_i$ // use of of CSI
6:         **for** $c \in candidates$ **do**
7:             **if** $c \notin result$ and $same\_order(c, combi, sim_\epsilon)$ **then**
8:                 $result \leftarrow result \cup \{c\}$
9:             **end if**
10:         **end for**
11:     **end for**
12:     **return** $result$
13: **end function**

The step to create combinations of cs points (line 3 of Algorithm 3) remains the same. On line 7, the point order check takes into account ϵ-similarity point matching by invoking $same\_order$ with sim_ϵ as the point matching function. It is important to note that the size of $\bar{\bar{p}}_i$ set introduced in Definition 4 depends on the selected ϵ similarity threshold. The lower the threshold, the more similar sequences will be retrieved. Our experiments will investigate the impact of ϵ on the result size, specifically the number of additional sequences.

6   Experimental Validation

This section reports on the experimental evaluation of our method and its comparison with the baseline. We start by describing the datasets used and outlining the experimental methodology for both exact sequence indexing and indexing based on point embeddings. Subsequently, we analyze and discuss the results obtained. All experiments were conducted on a Dell PowerEdge R440 server running Linux, equipped with 370 GB of RAM, and the implementation was carried out in Python. The index structure was implemented using a dictionary that maps each point (or pair of points) to a set of sequences.

For the embedding-based solution, we additionally rely on the multidimensional index provided by the W2V model, which retrieves points similar to a given point in a 10-dimensional vector space.

6.1 Data Preparation

We conduct our experiments on three datasets: Foursquare [27] and Gowalla [8] on New York city, and YFCC [23] on France. The aim is to demonstrate the effectiveness of our method across different datasets. Foursquare and Gowalla come with predefined points. YFCC is a larger dataset that contains geo-located check-ins but no POI information. To infer POIs from geolocations, we applied a grid-based partitioning of the space: each 100 square meter cell that contains some check-ins is considered as a POI, as proposed in [14].

In this study, we only consider points that have been visited more than 15 times. Points with low visit frequencies are excluded because they appear too rarely in the dataset. Consequently, sequences containing such rare points often lack sufficiently similar counterparts, preventing meaningful similarity comparisons. Our goal is to build sequences from frequently visited points in order to identify relevant neighbors that share common subsequences.

After selecting the points, we construct user sequences on a daily basis (from 0:00 am to 11:59 pm). Then, we keep only sequences with a realistic size, i.e. in [3,20]. More specifically, a sequence of two points is too short, However, our approach can be generalized since each point requires at least two neighboring points (before and after) to provide context. Therefore, we retain only sequences of length three or more. Sequences longer than 20 points are considered outliers that may introduce noise into the dataset and are thus excluded. Thus, we have 10,071 sequences with Foursquare, 23,145 sequences with YFCC, and 5,129 sequences with Gowalla.

We analyse the distribution of sequences sizes for the three datasets in Fig. 2, 3, and 4. The average sequence size is 5 for Foursquare, 6 for Gowalla, and 5 for YFCC. While our experiments are conducted on mobility datasets, the proposed method is applicable to other sequential data with moderate-length sequences.

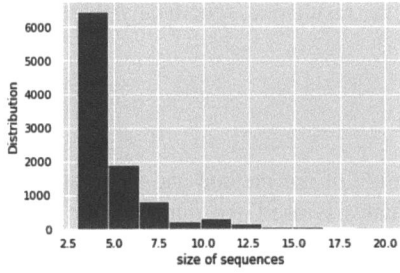
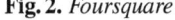
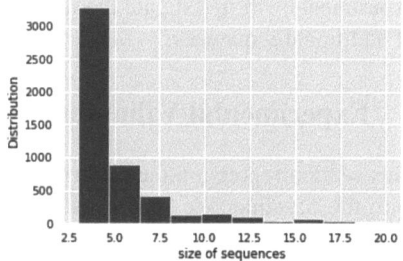

Fig. 2. *Foursquare*   **Fig. 3.** *Gowalla*

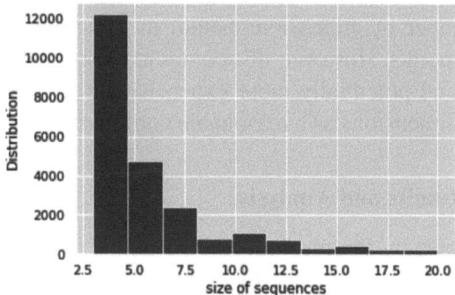

Fig. 4. YFCC

6.2 Exact Sequence Indexing: Experimental Methodology

For our experiments, we use the three datasets described in Sect. 6.1, and compare our proposed solution with the LCSS-based baseline. We use the sequences in the dataset as queries. For each query, we retrieve all the similar sequences, with various similarity thresholds ranging from 0.1 to 1, as well as the time taken to compute them. We focus on the effect of query size on response time, and report the average response times for each query size.

As outlined in Sect. 4.2, SISIS relies on the calculation of $C_{cs}^{|q|}$ combinations of cs points from q. As the number of combinations gets larger, SISIS performs slower. Therefore, we investigate the worst-case scenario where $C_{cs}^{|q|}$ is maximized, *i.e.* $cs = |q|/2$. Accordingly, we set the threshold S to 0.5 in all the following experiments.

The experimental methodology described here applies to indexing sequences by a single point (described in Sect. 4.2), and also to indexing sequences by pairs of points (described in Sect. 4.3). We carry out this experiment on Foursquare dataset in the Sect. 6.4, and subsequently compare the results with those obtained with the Gowalla and YFCC datasets in Sect. 6.4.

6.3 Sequence Indexing Based on Point Embeddings: Experimental Methodology

In this section, we perform experiments using the Foursquare dataset and compare the results with those obtained using SISIS (exact method) on the same dataset. The aim is to demonstrate that SISIS* which is based on embeddings produces more interesting results than SISIS, thanks to the contextual similarity. We trained W2V with the following parameters: vector_size = 10, epochs = 5, window = 5, and left all other parameters at their default values. The user threshold is still $S = 0.5$ and ϵ is in $\{0.65, 0.7, 0.75, 0.8, 0.85, 0.9, 0.95, 1\}$.

We selected a 10-dimensional vector space for training the Word2Vec model in SISIS* to strike a balance between semantic expressiveness and computational efficiency. A lower dimensional space (e.g., 10) is generally sufficient to capture basic contextual similarities between points, especially when the dataset is not extremely large or complex like ours. It also ensures that the computation of similarity and the construction of the embedding-based index remain fast.

Increasing the number of dimensions could, in principle, capture more nuanced relationships between points. However, this comes at the cost of higher memory usage, longer training time, and potentially more expensive similarity computations. Therefore, we opted for 10 dimensions as a reasonable compromise.

6.4 Experimental Results and Analysis

In this section, we report on the results obtained with exact indexing (using 1 and 2 points), and embedding-based indexing, following the experimental protocol described in Sect. 6.

SISIS Exact Indexing Using Single Points

Impact of Query Size on SISIS Execution Time. For exact indexing with a single point, we analyse the average execution time as a function of the query size specified by the user. Since the distribution of sequence sizes in the datasets is uneven (see Sect. 6.1), the number of queries per size also varies. We compute average response times for each query size, but do not normalize by the number of queries, which may affect the interpretation of the results for less frequent sizes, as their average execution time may be more sensitive to outliers.

This is one of the reasons why we choose in Sect. 6.1 to exclude infrequent sequences from the datasets. These sequences are typically very long (over 20 points), occur less than 6 times in the Foursquare dataset.

The threshold is fixed to 0.5 to represent the worst-case scenario for SISIS. The results are illustrated in Fig. 5 and 6.

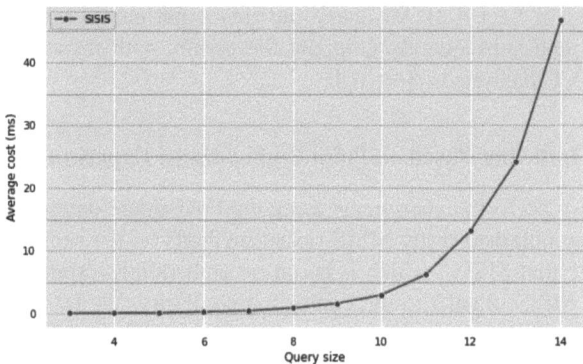

Fig. 5. SISIS response time for query sizes in the range [3, 14] - Foursquare dataset

First, we observe on Fig. 5 that SISIS response time is less than 1 ms for query size smaller than 8. Thus, SISIS meets users expectation in terms of response time.

On Fig. 6, we notice that our solution (SISIS in blue) is faster than LCSS-based baseline (in red) for all queries smaller than 17, above which size it becomes slower than

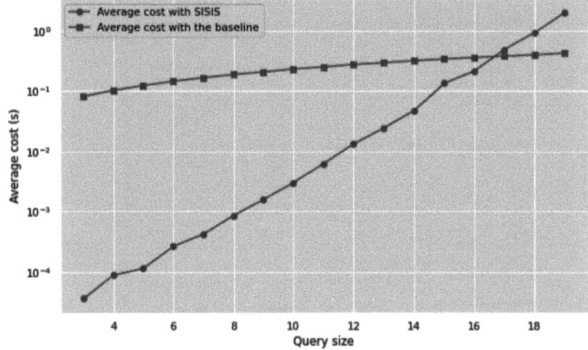

Fig. 6. Average time as a function of query size - Foursquare dataset (Color figure online)

LCSS-based baseline. However, in domains such as mobility, it is unlikely that users would formulate queries involving a large number of points, as this would not reflect realistic user behavior. Therefore, we can claim that SISIS largely outperforms the LCSS-based baseline for all realistic queries. Note that we also experimented SISIS and LCSS-based baseline for other threshold values (0.1 and 0.9). As expected (see Figs. 7 and 8), the results are more in favour of SISIS than for 0.5.

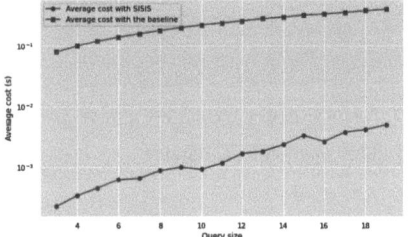

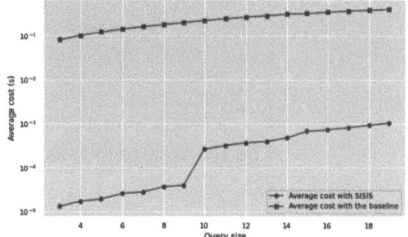

Fig. 7. *Average time as a function of query size (threshold = 0.1)*

Fig. 8. *Average time as a function of query size (threshold = 0.9)*

Generalization to other Datasets: Gowalla and YFCC. We performed the same experiment on other datasets to show that SISIS can produce good results on any other dataset. The chosen datasets are: Gowalla and YFCC. The Figs. 9 and 10 show the results obtained with the experiments conducted on these 2 datasets.

We observe that the results are comparable with the ones obtained for Foursquare, which confirms that SISIS outperforms LCSS-based baseline for reasonable query sizes and various datasets. We highlight the benefit of SISIS for queries of average sequence size (6 and 5 for Gowalla and YFCC respectively). SISIS is 330x (resp. 2200x) faster than LCSS-based baseline.

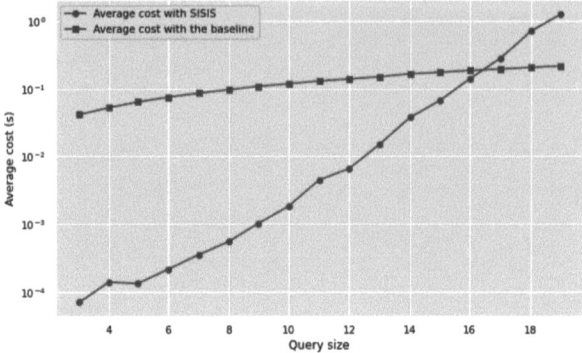

Fig. 9. Average time as a function of query size - Gowalla dataset

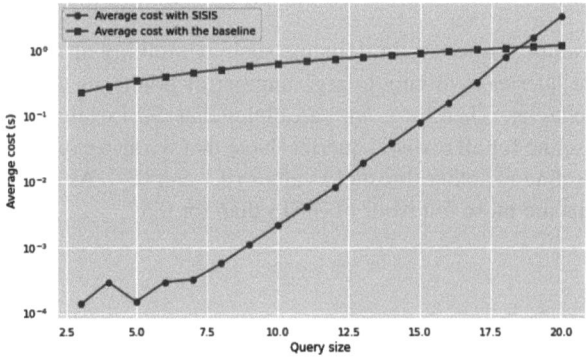

Fig. 10. Average time as a function of query size - YFCC dataset

SISIS: Estimated vs. Experimental cost on Foursquare Dataset. In Sect. 4.4, we computed the theoretical cost of SISIS. In this section, we compare the estimated cost with the cost obtained from experimental results on Foursquare dataset. We calculate the theoretical average cost (TC) as a function of query size and plot it alongside the experimental results.

As shown in Fig. 11, the estimated cost values (in green) closely match the experimental costs (in blue) obtained from experiments, with a Mean Absolute Error (MAE) of 0.14 and a Mean Squared Error (MSE) of 0.11. Since the estimated cost of SISIS closely match the experimental one, we can conclude that our implementation is correct. The theoretical estimate is an accurate reflection of reality.

Baseline Method: Estimated vs. Experimental Cost on Foursquare. We compare the theoretically estimated cost of the LCSS-based baseline with the experimental cost obtained from our experiments on Foursquare dataset. As done in Sect. 6.4, we set the maximum sequence size to $n = 20$. We then calculate the theoretical average cost (TC) as a function of query size and plot it alongside the experimental results.

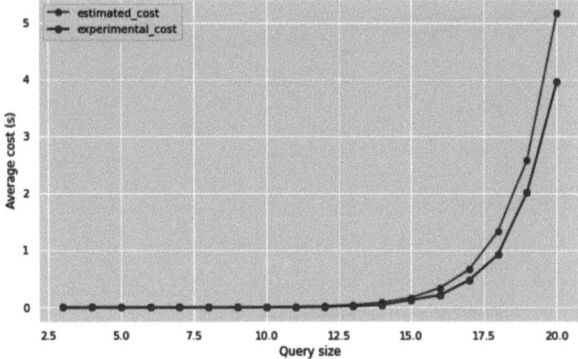

Fig. 11. SISIS estimated cost vs. experimental cost - Foursquare (Color figure online)

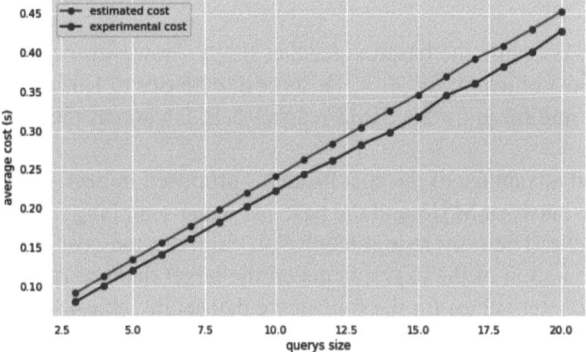

Fig. 12. Baseline estimate cost vs. experimental cost - Foursquare (Color figure online)

From Fig. 12, the estimated cost values (in green) closely match the experimental costs (in blue) obtained from experiments, with a Mean Absolute Error (MAE) of 0.02 and a Mean Squared Error (MSE) of 0.0004. The baseline cost estimate is in line with reality.

Validating the Cost-Based Choice Between SISIS and the Baseline. The aim of this discussion is to validate the hypothesis stated in Sect. 4.4: by applying the theoretical formulas derived earlier, and for small to medium query sizes, SISIS outperforms the baseline in terms of cost.

By comparing the estimated costs of both the baseline and SISIS, we obtain the results illustrated in Fig. 13. As expected, since the estimated costs for both the baseline and SISIS closely align with the real values observed in the experiments, the same trend emerges: SISIS remains more efficient than the baseline up to a certain query size, as early discussed. Figure 13 shows that for small and medium query values (up to 16), SISIS achieves lower costs compared to the baseline. For instance, when $|q| = 5$ the cost of SISIS is $0.0002s$ whereas the baseline costs $0.12s$. Similarly, for $|q| = 16$ SISIS

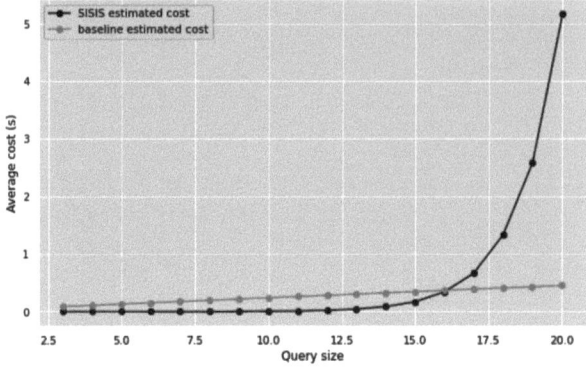

Fig. 13. Baseline estimated cost vs. SISIS estimated cost - Foursquare

has a cost of $0.34s$, while the baseline reaches $0.39s$. However, beyond this threshold, SISIS becomes less efficient: for $|q| = 18$, its cost increases to $1.32s$, compared to 0.44s for the baseline, and for $|q| = 20$, SISIS reaches $5.16s$, whereas the baseline remains at $0.49s$.

One practical advantage of the cost formulas proposed in Sect. 4.4 is their ability to guide the choice between SISIS and the baseline when searching for sequences similar to a given query q. The only requirement is to calibrate the cost formula parameters during the computation of the experiments on the target dataset. As an illustration, we provide the parameter values for the foursquare dataset in Table 4.

Table 4. Notations used for theoretical cost formulas, and their values

Notation	Description	Value
$|q|$	Query size	Varies according to input query size
cs	Combination size	$|q|/2$
iat	Intersection average computation time	$6.6e - 07$
fat	Filtering average computation time	$2.58e - 06$
as	Average size of candidate sequences	8.56
C	$t_{as} * a_t * m$	0.024

By applying the formulas with the values mentioned, we obtain the same results as those reported in the previous experiment.

SISIS Exact Indexing Using Point Pairs We compare the performance of the single point index (1P) and of the point-pair index (2P) using the Foursquare dataset. As shown in Fig. 14, the point-pair index (represented by the green curve) performs faster than the single point index (represented by the blue curve). On Fig. 15, the performance benefit ranges from 5x to 8x (with an average benefit of 6x) for queries of size 3 to 12.

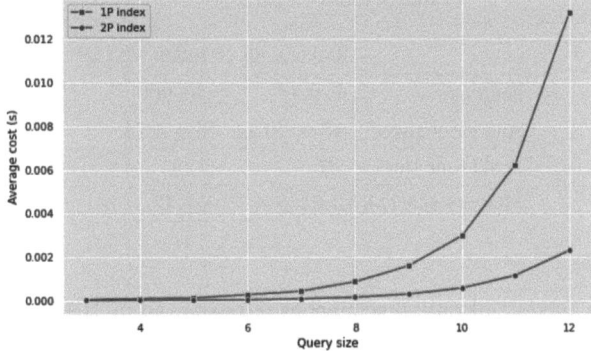

Fig. 14. 1 Point index vs. 2 Points index - Foursquare (Color figure online)

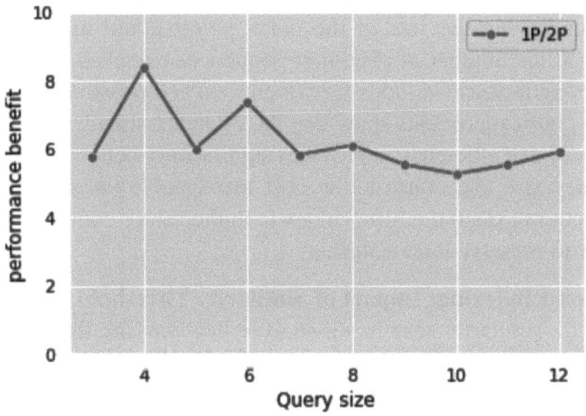

Fig. 15. Response time 1P/2P - Foursquare

As a counter part, the index creation time increases from 12 ms to 120 ms as reported in Table 5. Indeed, the 2P index contains 40,900 entries, compared to only 2,900 in the 1P index, making it approximately 14 times larger.

This moderate increase in the number of entries is due to the construction strategy used. For the 1P index, we extract all distinct points from the dataset sequences, resulting in 2,900 unique entries. In contrast, the 2P index is built by considering all possible pairs of these points. While this could theoretically lead to a quadratic number of combinations (i.e., in the millions), in practice, we retain only the pairs that actually occur together in at least one sequence in the dataset. This filtering significantly reduces the total number of entries, resulting in a manageable size of 40,900 entries, which remains tractable in real-world scenarios.

As anticipated, the 2P index demonstrates a high level of selectivity due to its reliance on pairs of points for indexing. It significantly narrows down the candidate set during the retrieval process. In our experiments on the Foursquare dataset, this selectiv-

Table 5. Index construction cost

	Index with 1P	Index with 2P
#entries	2, 900	40, 900
Avg #sequences	15 ± 20.24	3 ± 4.74
Build time (ms)	12	120
Memory size (Mo)	3.61	17.32

ity results in an average of only 3 sequences retrieved out of 10,000, highlighting the index's ability to effectively filter out non-relevant sequences.

SISIS currently supports indexing either single point (1P) or ordered pairs of points (2P). The choice of stopping at pairs, rather than considering more selective units such as points triples, is primarily motivated by the trade-off between selectivity and efficiency. While increasing the size of the index points could improve selectivity and potentially reduce the number of candidate sequences retrieved, it would come at the cost of a substantial increase in index size (# entries) and construction time.

Additionally, our experiments show that the 2P index already offers a good balance between selectivity and efficiency: it provides significantly better filtering than 1P, while keeping the index size and construction cost manageable (as shown in Table 5). As such, we opted not to extend the SISIS index to higher-order combination of points like triplets, in order to preserve its scalability.

Embedding-Based Indexing: Impact of Similarity Threshold and Cost Evaluation
In the context of ϵ-similarity search, we analyze how varying the relaxation threshold ϵ affects the number of sequences retrieved, using the Foursquare dataset. Recall that SISIS* extends the standard SISIS method, by integrating points embeddings to capture contextual similarity between sequences. This embedding-based approach allows the retrieval of a broader set of sequences, including those that do not strictly match the query points but that contain semantically related ones.

On Fig. 16, we vary the value of the similarity threshold ϵ on the x-axis. The y-axis is the average percentage of extra sequences retrieved by the queries. For example, for $\epsilon = 0.72$, we retrieve twice as many sequences as SISIS. This increase is due to the fact that the number of neighbors increases with ϵ. We illustrate this on Fig. 17. Specifically, we obtained 14 neighbors for $\epsilon = 0.72$ and only 1 neighbor for $\epsilon \geq 0.9$.

Moreover, the cost remains close to that of the standard SISIS algorithm: less than 0.2 s for $\epsilon > 0.72$ (see Fig. 18), demonstrating the efficiency of our approach. For $\epsilon < 0.72$ the cost actually increases significantly (up to 1.35 s for $\epsilon = 0.65$), but such a range of ϵ values would rarely be used in practice because it returns too many sequences (more than twice as many as with the standard SISIS method).

Another aspect that our experiments aimed to investigate concerns the relevance of embeddings. To this end, we reduced the embeddings dimension to 2 (instead of the previous dimension value of 10), so that we can plot them on Fig. 19. We observed that they are rather uniformly dispersed.

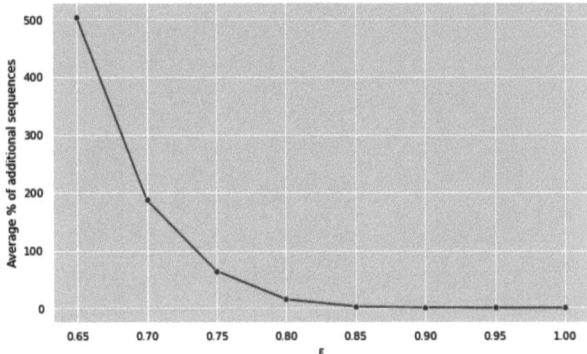

Fig. 16. Effect of ϵ on the average percentage of additional sequences - Foursquare

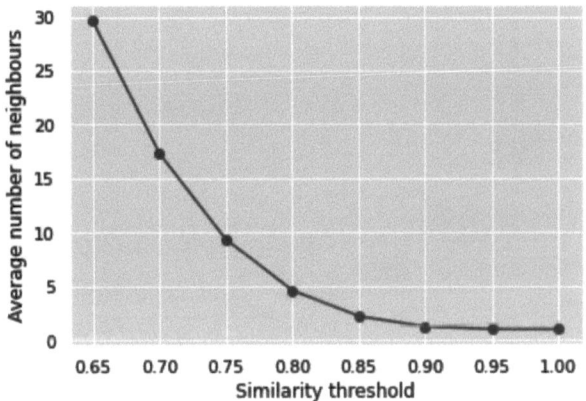

Fig. 17. Average number of neighbors per similarity threshold - Foursquare

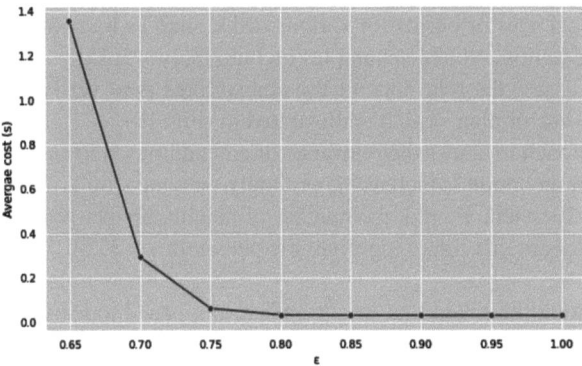

Fig. 18. average cost per similarity threshold - Foursquare

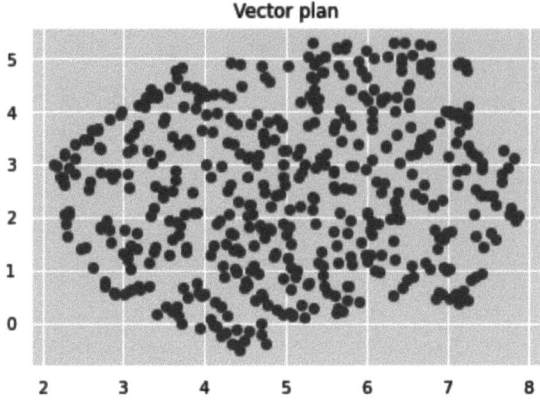

Fig. 19. 500 points sample (in 2D plan)

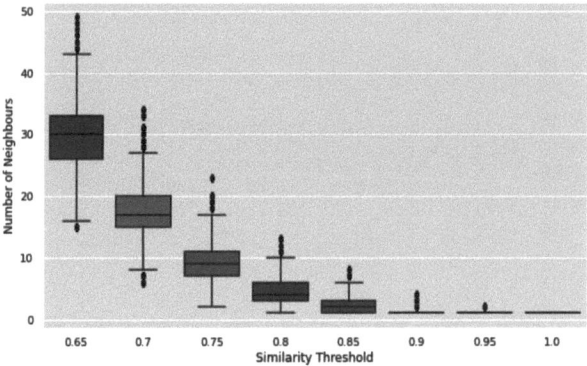

Fig. 20. #Neighbours per point

Several recent works have demonstrated the efficacy of using embeddings to represent mobility sequences or points for various tasks, such as location classification and next location prediction. Notable examples include [3, 14, 15, 25].

One way to assess the relevance of the embeddings is to verify that they are well dispersed on the vector plan, which is illustrated in (Fig. 19).

Another approach to assess the relevance of embeddings is to show that the number of neighbours for any point is increasing gradually with the similarity threshold. This is the case on Fig. 20 where we plot the number of neighbours per point according to the threshold (boxplots in this figure highlight the percentiles 5,25,50,75,95 of the number of neighbours).

Finally, it should be noted that in our embedding-based indexing experiments, we focus exclusively on the 1P index and do not include the 2P index. The 2P index, which involves sequence associations for all possible point pairs (p_i, p_j) and the identification of ϵ-similar pairs (p_k, p_l), results in a significantly higher number of entries and a substantially greater index construction cost. Specifically, the number of entries increases

Table 6. Embeddings indexes construction cost

	Index with 1P	Index with 2P
#entries	1, 754	8, 645
Build time (s)	240	2, 220

from 1,754 in the 1P index to 8,645 in the 2P index (almost 5x larger), and the construction time rises from 240 s to 2,220 s (almost 10x slower), as shown in Table 6.

The significant increase in index construction time, as shown in Table 6 compared to the values reported in Table 5 is explained by the use of point embeddings. In this setting, the index is built not only from the exact points in the dataset, but also from all points that are considered similar in the embedding space.

For the 1P embedding-based index, we retrieve sequences containing not only a given point, but also those that contain any point deemed similar to it. This results in a broader search and more sequences to process during index construction.

The 2P embedding-based index further increases this complexity. For each point pair, we retrieve not only the sequences containing the exact pair but also those matching any combination involving similar points, whether the first point, the second point, or both are replaced by similar ones. This leads to a combinatorial increase in the number of candidate sequences to process, which significantly increases the construction time compared to the exact indexing.

7   Conclusion

This study demonstrates the effectiveness of an indexing-based approach for sequence similarity search when compared to a baseline method that relies on the Longest Common Subsequence (LCSS) algorithm. To address the performance limitations inherent in LCSS-based search, we introduce *SISIS* (Sequence Indexing for Similarity Search). SISIS employs an exact indexing approach for sequence similarity search. By leveraging two distinct sequence indexing techniques, our approach consistently produces results identical to the LCSS-based baseline, while offering significantly improved performance for queries of a reasonable size.

Nevertheless, SISIS may be overly restrictive, as it relies on exact matching of points, which can result in too few (or even no) returned results. To mitigate this limitation, we have developed and implemented *SISIS**, a relaxed variant of SISIS. This extension enables more comprehensive retrieval of similar sequences by accounting for semantic similarities between points, rather than relying solely on exact matches. The relaxed method yields a significantly higher number of results (approximately twice as many as the exact method with a similarity threshold of 0.72) while preserving computational efficiency.

We conducted experiments on three real-world datasets, demonstrating that the proposed approach significantly outperforms the baseline up to a query size of 17 on the Foursquare dataset. However, as the query size increases beyond this threshold, its computational cost exceeds that of the LCSS-based baseline, showing that SISIS is not suitable for very large queries.

Currently, *SISIS* is limited to optimising only the LCSS-based similarity measure, as it retrieves sequences based on exact point matches and sequential order. This definition does not directly apply to other similarity measures that rely on different assumptions.

For instance, DTW (Dynamic Time Warping) is designed to compare continuous time series by aligning them through temporal warping. It measures similarity by computing the minimum cumulative Euclidean distance between points in the two sequences, even if they are not aligned in time. This flexibility allows DTW to handle shifts and stretching in the temporal dimension, which SISIS is not designed to support.

Likewise, ERP (Edit distance with Real Penalty) is based on the concept of edit distance. It computes the cost of transforming one sequence into another using insertions, deletions, and substitutions, where substitution costs are calculated via the Euclidean distance between points. Unlike LCSS, which focuses on counting matching points, ERP considers the accumulated numerical dissimilarity between sequences.

Because SISIS relies entirely on LCSS and its discrete matching strategy, it cannot currently support similarity measures such as DTW or ERP. Extending SISIS into a more general framework that accommodates different types of similarity metrics remains an open and promising direction for future work.

Another avenue we could explore is extending our approach to domains beyond mobility trajectories, thereby showcasing its applicability to a broader spectrum of sequential data. While the proposed method is structurally generic, its deployment in other contexts entails specific challenges. For instance, in the case of textual data, we may face a dramatic growth in index size due to the vast number of distinct words. Similarly, when applied to biological DNA sequences, differences in sequence length relative to mobility data may impact the overall effectiveness of the approach. Addressing such challenges will thus necessitate dedicated algorithmic adaptations

References

1. Altschul, S.F., Gish, W., Miller, W., Myers, E.W., Lipman, D.J.: Basic local alignment search tool. J. Mol. Biol. **215**(3), 403–410 (1990). https://doi.org/10.1016/S0022-2836(05)80360-2. https://www.sciencedirect.com/science/article/pii/S0022283605803602
2. Berndt, D.J., Clifford, J.: Using dynamic time warping to find patterns in time series. In: Fayyad, U.M., Uthurusamy, R. (eds.) Knowledge Discovery in Databases: Papers from the 1994 AAAI Workshop, Seattle, Washington, USA, July 1994. Technical Report WS-94-03, pp. 359–370. AAAI Press (1994). https://dblp.org/rec/conf/kdd/BerndtC94.bib
3. Biester, L., Banea, C., Mihalcea, R.: Building location embeddings from physical trajectories and textual representations. In: Wong, K., Knight, K., Wu, H. (eds.) Proceedings of the 1st Conference of the Asia-Pacific Chapter of the Association for Computational Linguistics and the 10th International Joint Conference on Natural Language Processing, AACL/IJCNLP 2020, Suzhou, China, 4–7 December 2020, pp. 425–434. Association for Computational Linguistics (2020). https://doi.org/10.18653/V1/2020.AACL-MAIN.44
4. Bringmann, K., Cohen-Addad, V., Das, D.: A linear-time $n^{0.4}$-approximation for longest common subsequence. ACM Trans. Algorithms **19**(1), 9:1–9:24 (2023). https://doi.org/10.1145/3568398

5. Chan, W., Zhang, Y., Fung, S.P.Y., Ye, D., Zhu, H.: Efficient algorithms for finding a longest common increasing subsequence. J. Comb. Optim. **13**(3), 277–288 (2007). https://doi.org/10.1007/S10878-006-9031-7

6. Chen, L., Ng, R.T.: On the marriage of LP-norms and edit distance. In: Nascimento, M.A., Özsu, M.T., Kossmann, D., Miller, R.J., Blakeley, J.A., Schiefer, K.B. (eds.) (e)Proceedings of the Thirtieth International Conference on Very Large Data Bases, VLDB 2004, Toronto, Canada, August 31 - September 3 2004, pp. 792–803. Morgan Kaufmann (2004). https://doi.org/10.1016/B978-012088469-8.50070-X. http://www.vldb.org/conf/2004/RS21P2.PDF

7. Chen, L., Özsu, M.T., Oria, V.: Robust and fast similarity search for moving object trajectories. In: Özcan, F. (ed.) Proceedings of the ACM SIGMOD International Conference on Management of Data, Baltimore, Maryland, USA, 14–16 June 2005, pp. 491–502. ACM (2005). https://doi.org/10.1145/1066157.1066213

8. Cho, E., Myers, S.A., Leskovec, J.: Friendship and mobility: user movement in location-based social networks. In: Apté, C., Ghosh, J., Smyth, P. (eds.) Proceedings of the 17th ACM SIGKDD International Conference on Knowledge Discovery and Data Mining, San Diego, CA, USA, 21–24 August 2011, pp. 1082–1090. ACM (2011). https://doi.org/10.1145/2020408.2020579

9. Gotoh, O.: An improved algorithm for matching biological sequences. J. Mol. Biol. **162**(3), 705–708 (1982). https://doi.org/10.1016/0022-2836(82)90398-9. https://www.sciencedirect.com/science/article/pii/0022283682903989

10. Hajiaghayi, M., Seddighin, M., Seddighin, S., Sun, X.: Approximating LCS in linear time: beating the $\sqrt{n}$ barrier. CoRR **abs/2003.07285** (2020). https://arxiv.org/abs/2003.07285

11. Han, T.S., Ko, S.-K., Kang, J.: Efficient subsequence matching using the longest common subsequence with a dual match index. In: Perner, P. (ed.) MLDM 2007. LNCS (LNAI), vol. 4571, pp. 585–600. Springer, Heidelberg (2007). https://doi.org/10.1007/978-3-540-73499-4_44

12. Hirschberg, D.S.: A linear space algorithm for computing maximal common subsequences. Commun. ACM **18**(6), 341–343 (1975). https://doi.org/10.1145/360825.360861

13. Hirschberg, D.S.: Algorithms for the longest common subsequence problem. J. ACM **24**(4), 664–675 (1977). https://doi.org/10.1145/322033.322044

14. Jarrad, S., Naacke, H., Gançarski, S., Gueye, M.: Embedding-enhanced similarity metrics for next POI recommendation. In: Gusikhin, O., Hammoudi, S., Cuzzocrea, A. (eds.) Proceedings of the 12th International Conference on Data Science, Technology and Applications, DATA 2023, Rome, Italy, 11–13 July 2023, pp. 247–254. SCITEPRESS (2023). https://doi.org/10.5220/0012060300003541

15. Lin, Y., Wan, H., Guo, S., Lin, Y.: Pre-training context and time aware location embeddings from spatial-temporal trajectories for user next location prediction, pp. 4241–4248 (2021). https://doi.org/10.1609/AAAI.V35I5.16548

16. Magdy, N., Sakr, M.A., Mostafa, T., El-Bahnasy, K.: Review on trajectory similarity measures. In: 2015 IEEE Seventh International Conference on Intelligent Computing and Information Systems (ICICIS), pp. 613–619 (2015). https://doi.org/10.1109/IntelCIS.2015.7397286

17. Mikolov, T., Chen, K., Corrado, G., Dean, J.: Efficient estimation of word representations in vector space. In: Bengio, Y., LeCun, Y. (eds.) 1st International Conference on Learning Representations, ICLR 2013, Scottsdale, Arizona, USA, 2–4 May 2013, Workshop Track Proceedings (2013). http://arxiv.org/abs/1301.3781

18. Needleman, S.B., Wunsch, C.D.: A general method applicable to the search for similarities in the amino acid sequence of two proteins. J. Mol. Biol. **48**(3), 443–453 (1970). https://doi.org/10.1016/0022-2836(70)90057-4. https://www.sciencedirect.com/science/article/pii/0022283670900574

19. Notredame, C., Higgins, D.G., Heringa, J.: T-coffee: a novel method for fast and accurate multiple sequence alignment11edited by J. Thornton. J. Mol. Biol. **302**(1), 205–217 (2000). https://doi.org/10.1006/jmbi.2000.4042. https://www.sciencedirect.com/science/article/pii/S0022283600940427

20. Sellers, P.H.: An algorithm for the distance between two finite sequences. J. Comb. Theory A **16**(2), 253–258 (1974). https://doi.org/10.1016/0097-3165(74)90050-8

21. Smith, T., Waterman, M.: Identification of common molecular subsequences. J. Mol. Biol. **147**(1), 195–197 (1981). https://doi.org/10.1016/0022-2836(81)90087-5. https://www.sciencedirect.com/science/article/pii/0022283681900875

22. Su, H., Liu, S., Zheng, B., Zhou, X., Zheng, K.: A survey of trajectory distance measures and performance evaluation. VLDB J. **29**(1), 3–32 (2020). https://doi.org/10.1007/S00778-019-00574-9

23. Thomee, B., et al.: YFCC100M: the new data in multimedia research. Commun. ACM **59**(2), 64–73 (2016). https://doi.org/10.1145/2812802

24. Toohey, K., Duckham, M.: Trajectory similarity measures. ACM SIGSPATIAL Special **7**(1), 43–50 (2015). https://doi.org/10.1145/2782759.2782767

25. Wan, H., Lin, Y., Guo, S., Lin, Y.: Pre-training time-aware location embeddings from spatial-temporal trajectories. IEEE Trans. Knowl. Data Eng. **34**(11), 5510–5523 (2022). https://doi.org/10.1109/TKDE.2021.3057875

26. Wang, H., Su, H., Zheng, K., Sadiq, S., Zhou, X.: An effectiveness study on trajectory similarity measures. In: Wang, H., Zhang, R. (eds.) Twenty-Fourth Australasian Database Conference, ADC 2013, Adelaide, Australia, February 2013. CRPIT, vol. 137, pp. 13–22. Australian Computer Society (2013)

27. Yang, D., Zhang, D., Zheng, V.W., Yu, Z.: Modeling user activity preference by leveraging user spatial temporal characteristics in LBSNs. IEEE Trans. Syst. Man Cybern. Syst. **45**(1), 129–142 (2015). https://doi.org/10.1109/TSMC.2014.2327053

28. Yi, B., Jagadish, H.V., Faloutsos, C.: Efficient retrieval of similar time sequences under time warping, pp. 201–208 (1998). https://doi.org/10.1109/ICDE.1998.655778

A Categorical Representation of Multi-model Data to Prevent Data Migration Mismatch

Annabelle Gillet$^{(\boxtimes)}$ and Éric Leclercq

Laboratoire d'Informatique de Bourgogne - UR 7534, Université Bourgogne Europe,
Dijon, France
{annabelle.gillet,eric.leclercq}@ube.fr

Abstract. To cope with the increasing diversity of data uses, storage systems have evolved towards multi-model systems. Indeed, as each data model has its own characteristics, constraints and operators, they can be best suited for a specific use case but not for another. Migrations between models are therefore essential to respond to the multiplicity of use cases, but are complex to manage because of the differences among models. As such, some constraints might not be preserved during a migration process, and some others might have to be created. Furthermore, the impact of a migration depends on the data schema: if the schema does not use a constraint of its source model, it is unimportant that the destination model does not support this constraint. Thus, it is necessary to specify a formal framework able to represent data models and their constraints, data schemas and data migrations. We rely on category theory to propose such framework. It uses categories to represent schemas and models, specific constructs such as isomorphisms, products and pullbacks to represent constraints, and functors to represent migrations. These elements serve to assess the impact of a migration on a data schema, depending on the source and destination models. Based on this framework, we define categories for the relational, JSON and property graph data models. We propose a prototype application implementing the formal framework. It can connect to data sources to automatically build the corresponding categorical schema, and display the impacts of a migration to users. An experimental evaluation of the execution time demonstrates the applicability of the approach.

Keywords: Multi-model data · Data migration · Category theory

1 Introduction

Extracting value from data is the ultimate goal of data management. This task is not straightforward, as some characteristics of data hinder their processing [1]. Among these characteristics, variety is of major concern. Multiplication of data sources increases the heterogeneity of data regarding their content, their quality,

© The Author(s), under exclusive license to Springer-Verlag GmbH, DE, part of Springer Nature 2026
A. Hameurlain et al. (Eds.): *Transactions on Large-Scale Data- and Knowledge-Centered Systems LIX*, LNCS 16240, pp. 61–93, 2026.
https://doi.org/10.1007/978-3-662-72449-1_3

and their model. However, as stated by M. Stonebraker [40] "One size does not fit all", and to manage this variety, numerous data management systems have emerged, each providing specific data models and sets of operators to query data and fit particular use cases.

To cope with multi-model data, storage systems have evolved towards multi-model systems. Some database management systems (DBMSs) propose several models at the same time [32], but they cannot be easily extended to add new models. Other systems such as polystores [43], data lakes [22] or data spaces [23] aim to benefit from each model and their specific operators individually (e.g., path finding queries in the graph model) by using jointly different DBMSs. They can integrate new data models more easily by adding a different DBMS, but at the cost of an increased complexity.

In these systems, data migrations are essential to satisfy multiple use cases. When ingesting data, the source model might not be the most suitable model for manipulating data, and it might require to migrate data towards a system supporting another data model. For example, social network data are usually supplied in the JSON format, that can be transformed into a graph model to study the interaction among users, into a relational model to store the main characteristics of messages such as the hashtags or the publication time, and into a textual model to study the content of publications [16,21]. Once data are stored, they can be used for querying or for further data processing. It can result in the application of an operator not available in the current system, thus requiring a migration towards another system, or in the execution of an algorithm working on data originating from multiple sources, that must be integrated into a common model suitable for the algorithm [24,30]. Query optimization in multi-model systems may also lead to data migration. As some models are best tailored to execute some types of query, the performances can be enhanced by migrating data from its source DBMS to a DBMS that can better handle the query, even when the migration time is included [4,36]. Optimizations can also be applied with operators processing data of multiple systems, as for example with the bind join between a large and a small datasets [28], that sends the small dataset into the system of the large dataset to filter it and to avoid retrieving unnecessary data.

However, in multi-model systems, data migration is a daunting process [33]. Indeed, transforming data from a model to another can be done in multiple ways, and requires to add constraints (e.g., adding a foreign key constraint stating that the attribute of the referencing table must be equal to a primary key attribute of the referenced table when migrating from the graph model to the relational model) or to lose expressivity (e.g., loosing a primary key constraint when migrating from the relational model to the JSON model). The acceptability of these alterations depends on the context of the migration. One may want to use an equivalent model if the goal is to change the system used for storing data, but one could accept some losses regarding the constraints of the data if the goal is just to execute an analysis algorithm. Furthermore, a transformation depends on the schema of the data (e.g., if there is no primary key constraint applied on data stored with the relational model, it is not important that the JSON model does not support such constraint). Thus, as data migrations can be costly to process and can lead to errors or data inconsistencies [3], it is essential to identify beforehand the constraints that must be dropped or created to perform safely the migration.

In this article, we propose an extensible formal framework which relies on category theory to assess the impacts of a data migration on a schema. To do so, we represent data models and data schemas with categories, as well as constraints with different constructs (such as isomorphisms, products or pullbacks). Functors are used to ensure that a schema complies to its model, and to verify which constraints are lost during the migration and which constraints need to be created to comply to the destination model. We propose an implementation of the formal framework in an open-source prototype, that is able to extract schemas from data sources, use predefined migrations or let users define their specific migrations, and automatically assess the impact of a migration on a schema.

The rest of the article is organized as follows. Section 2 presents related works on multi-model data management and on the use of category theory for data management. Section 3 is an introduction to category theory detailing key notions to formalize multi-model data, Sect. 4 defines how to use category theory to represent data models and their constraints, and shows how to define a data model through examples for the JSON, property graph and relational models, Sect. 5 introduces categorical data schema, that are categories linked to their model with a functor ensuring that the schema complies to its model, and presents 3 example schemas, Sect. 6 demonstrates how to represent migrations with functors and how to assess their impacts on a schema. This mechanism is illustrated by applying migrations on the example schemas of the previous section. Section 7 presents the prototype and an experimental evaluation of its execution time. Finally Sect. 8 concludes the article with some perspectives.

2   Related Work and Motivations

Following research works on model management initiated twenty years ago [9–12], the management of multi-model data has been addressed by two main approaches: specifying mappings among data models and schemas, or using a pivot model able to abstract multiple data models. Among these works, category theory has been experienced, mainly as a formalism to express schemas regardless of their model. These works led us to consider category theory as a formal approach for unifying the management of multi-model data while remaining flexible enough to fit user needs.

2.1   Multi-model Data Management

Bernstein et al. proposed a data model specifying models, schemas and mappings [12]. The data model uses formal object oriented structures and proposes two main abstractions: 1) model, which captures the structure of information artifacts, such as database schemas, XML structures, interface definitions, complex document structures, or semantic networks; and 2) mapping, which captures relationships between models such as transformations and matching. The formal data model is completed with an algebra that provides some operations such as matching, merging or mapping composition. By defining such set of operations, the authors aim to reduce the amount of code required for applications which manipulate multiple models and mappings.

Generic Role based Metamodel (GeRoMe) [26] is a model management system that supports definitions for models, mappings and operators. It gives roles to elements, that allow to check if an element has certain properties. The roles include Aggregate, Attribute or Type for example.

The BigDAWG polystore [18] allows to gather storage systems into islands (for example a relational or an array island) that each uses its own data model and query language. Adapters translate the query from the island language to the storage system language. A common interface is used to orchestrate the polystore, including components to plan, optimize and execute queries, as well as to migrate data between systems [47].

Darwin [41] is a middleware between a NoSQL database and a Java application, that can handle schema evolutions and data migrations between two versions of a same schema. In [14], this work has been pushed further to support migration between heterogeneous data stores, by relying on a metamodel that combines elements from the columnar, document, key-value and graph models.

Other proposals rely on a single pivot model to abstract multiple data models. Pivot models provide a single interface that can be used to query storage systems, regardless of their underlying data model. ESTOCADA [5] is based on a relational pivot model with query rewriting in presence of constraints. D4M [19] uses associative arrays to abstract relational and some NoSQL DBMSs. Cloud-MdsQL [27] is based on a relational model and can take advantage of specific functionalities of data stores by allowing local native queries called as functions. Abstra [7] uses a graph representation for abstracting CSV, relational, JSON, XML or property graph datasets.

An alternative to manage multi-model data is to provide a unified schema representation regardless of the model. As discussed in [9], it can be done with a directed graph with vertices representing the elements of the schemas, and edges representing mappings between elements of schemas. Using a labeled graph allows to reach more complex and detailed representations, by giving types to elements and mappings with labels [15], but are still limited (i.e., difficulties to represent equivalences or multiple levels of abstraction).

Most of approaches do not include directly data models in their representation, although they carry a lot of information. However, the diversity of data models requires flexibility in the representation to fully integrate all their constraints and specificities. There are some essential concepts to take into consideration, and category theory proposes constructs that can efficiently represent these concepts. First, multiple levels of abstraction are needed to represent data models and data schemas, and relationships are needed to link a schema to its model, and models among them. With category theory, the levels of abstraction can be defined in categories, and the navigation among these levels is ensured by functors. Second, constraints are diverse, and some of them are a subrepresentation of another constraint. It is essential to be able to represent all of them, and to define equivalences between constraints of different data models or schemas. In most representations, it cannot easily be defined without relying on external reasoning or validation mechanisms [39]. Category theory proposes natively several specific morphisms and constructs (such as isomorphisms and pullbacks), as well as equalities of morphisms (including equalities of compositions) with commutative diagrams, on which we can rely to represent constraints and equivalences.

2.2   Category Theory for Data Management

The use of category theory for database formalization has started in the nineties. Johnson and Dampney in [25] show how commutative diagrams from category theory have been used to define methodologies for ER-modelling with constraint specifications and process definition. The Functorial Data Model [34] relies on category theory to define a model close to object-oriented database models. It represents classes and objects with categories, relationships with pullbacks, inheritances with coproducts and views with subcategories. Diskin and Kadish [17], by studying works on category theory used in database theory, show that while there are precise descriptions of what is a relational schema or a semantic schema, there is no work on mapping between schemas expressed in other models such as ER or OO.

The works of Alagič and Bernstein [2,8] in the field of data integration use categories as a universal representation of schemas. Schema categories and morphisms carry structural and semantic properties. They apply transformations on both the schema structure and its integrity constraints. They also show how to address problems such as schema equivalence and schema integration.

In these early works, category theory was already identified as having all the requirements to deal with data regardless of their model. However, the representation of the model itself was not considered, and the theory was rather used to represent schemas with a unique formalism. More recent works continue to follow this direction.

Spivak et al. propose CQL, that uses category theory to represent relational schemas [35,37,38], and relies on morphisms to represent attributes, including primary keys and foreign keys. With functors, they define projection, union and join operators through the Functorial Query Language [46].

In [31], the authors focus on relational and JSON schemas. They propose to use right and left pushforward functors between schema categories to transform a relational schema into a JSON one through an inner or an outer join.

Uotila et al. [44] represent relational, graph, and hierarchical schemas and instances as categories. They use Kan lifts to validate at the same time the schema and the data transformations. They propose an implementation named MultiCategory [45].

In [29, 42], the authors propose a unified representation of multi-model data using category theory. The representation relies on two main categories: the `Schema` and the `Instance` categories, linked by functors. The proposition consists in an abstract model allowing to embed all the other models.

In all of the cited works, models are not directly represented in the formalization: the schema of data is the higher level of abstraction, and the transformations occur among schemas. However, including data models into the formalization is essential, as they carry important information, such as the structure that a schema must comply to. With data models integrated into the formalization, data migrations can be defined as relationships between a schema and its destination model, on which we could rely to assess the impact of a migration on a given schema (i.e., which constraints are lost and which constraints need to be created to comply to the destination model).

3   Introduction to Category Theory

In category theory, a category defines a structure with objects and their relationships, either through direct relations with morphisms or through compositions.

Definition 1 (Category). *A **category** C is composed of four basic elements:*

1. *$Ob(C)$, a collection of **objects**;*
2. *for each pair $x, y \in Ob(C)$, a set $Hom_C(x, y)$ representing **morphisms** from x to y, namely a means to map an object x (the domain) to an object y (the codomain). A morphism f from x to y is noted $f : x \to y$. We note Hom_C the set of all morphisms of C;*
3. *for each $x \in Ob(C)$, a particular morphism $id_x : x \to x$ known as the **identity morphism** on x, that acts as a unit operation;*
4. *for each triple $x, y, z \in Ob(C)$ and their morphisms, a **composition** $\circ$: $Hom_C(y, z) \times Hom_C(x, y) \to Hom_C(x, z)$. For two morphisms $f : x \to y$ and $g : y \to z$, the composition is noted $g \circ f : x \to z$.*

And of two laws:

1. *for a morphism $f : x \to y$ with $x, y \in Ob(C)$, we have $f \circ id_x = f$ and $id_y \circ f = f$;*
2. *for $f : w \to x$, $g : x \to y$ and $h : y \to z$ with $w, x, y, z \in Ob(C)$, we have $(h \circ g) \circ f = h \circ (g \circ f) \in Hom_C(w, z)$.*

The objects of a category are a rather abstract notion. They are not defined by *what* they are, but only by *how they relate* to other objects, thus making morphisms the central elements of a category. A category can represent elements of a structure with objects, and the relationships among these elements with morphisms, as it will be shown for data models in Sect. 4.1 and for data schemas in Sect. 5.

In figures, we will represent categories by boxes, objects by their name written in blue and morphisms by an arrow. Although identity morphisms exist for all objects, they are not represented to not overload figures.

Starting from categories, category theory provides useful constructions and mechanisms allowing to define different abstraction levels and navigate among them, to define properties directly in the structure of a category, and to preserve the structure of a category into another category.

3.1 Navigation Among Abstraction Levels

Categories can be mapped by functors. This mapping defines how each object and morphism of the source category relate to the destination category.

Definition 2 (Functor). *A **functor** F maps a category C to a category C'. It is noted $F : C \rightarrow C'$, and affects:*

1. *objects: for each object $x \in Ob(C)$ we have $F(x) \in Ob(C')$;*
2. *morphisms: for each pair of objects $x, y \in Ob(C)$ we have $F : Hom_C(x, y) \rightarrow Hom_{C'}(F(x), F(y))$, thus for a morphism $f \in Hom_C(x, y)$, we have $F(f) \in Hom_{C'}(F(x), F(y))$.*

To be valid, a functor must observe two laws:

1. *the preservation of identities: $\forall x \in Ob(C)$, $F(id_x) = id_{F(x)}$;*
2. *the preservation of compositions: for any triple $x, y, z \in Ob(C)$ with morphisms $g : x \rightarrow y$, $h : y \rightarrow z$, we have $F(h \circ g) = F(h) \circ F(g)$.*

As categories can represent various structures at different levels of abstraction, functors can describe how a level of abstraction relates to another. For example, a data model is a higher level of abstraction than a data schema, and both can be represented in a category, with a functor mapping the schema category to the model category.

In Sect. 4.5 we will see how to rely on functors to map a specific use of rules of a data model to its main data model, and how to map a schema to its data model in Sect. 5.

In the remainder of this article, functors are given in tables, in which the first column gives objects and morphisms in the source category, and the second column gives the result of the application of the functor in the destination category.

3.2 Properties are Carried by the Structure

A category can directly embed properties through three main mechanisms: commutative diagrams, specific morphisms and limits (or colimits). Each mechanism brings its own advantages.

Commutative diagrams define equalities among morphisms.

Definition 3 (Commutative diagram). *A **commutative diagram** is a part of a category in which all parallel morphisms (i.e., morphisms having the same domain and codomain) are equal.*

Commutative diagrams represent properties stating that relying on different relationships between two elements yields the same result. It can be useful to enforce some types of constraints in data model, for example to specify that a foreign key attribute in a referencing table is equal to a primary key attribute in the referenced table (as it will be shown in Sect. 4.4).

As morphisms are central elements of category theory, they can have different nature depending on their properties. For example, there exist monomorphisms, that are left-cancellable morphisms, epimorphisms, that are right-cancellable morphisms, or isomorphisms, that define equivalences among objects. In this article, we will show the usefulness of specific morphisms with isomorphisms.

Definition 4 (Isomorphism). *A morphism $f : x \to y$ is an **isomorphism** if it is invertible, namely there exists a morphism $f^{-1} : y \to x$ such that $f^{-1} \circ f = id_x$ and $f \circ f^{-1} = id_y$. Isomorphisms are noted $f : x \xdashrightarrow{} y$, with the dashed arrow representing the inverse morphism f^{-1}.*

Equivalence is an interesting notion, different than equality, allowing to state that two elements have the same properties without being the same entity. In data representation, this is useful for example to define identifiers, that must all refer uniquely to a tuple of data without being the same element (see Sect. 4.4).

Limits are a flexible mechanism to embed properties into a category. It relies on a pattern defined in an index category, mapped to the category in which to embed the property with two functors. A limit is a cone equipped with the universal property (see Fig. 1).

Definition 5 (Cone). *In a category $\mathcal{C}$, a **cone** is given with an index category I and two functors $J : I \to \mathcal{C}$; and $\Delta(v) : I \to \mathcal{C}$ a constant functor towards the object $v \in Ob(\mathcal{C})$ (i.e., a functor mapping every object in I to $v \in Ob(\mathcal{C})$ and every morphism in I to $id_v \in Hom_{\mathcal{C}}(v, v)$). The object v is called the vertex of the cone, and the base B contains objects and morphisms of the category $\mathcal{C}$ that are mapped by the functor J. For every object $o \in Ob(I)$, there must exist a morphism $\rho : v \to J(o)$ called a component of the cone, and for each morphism $m \in Hom_I$, the diagram composed of $J(m) : b_1 \to b_2$ and the components $\rho_1 : v \to b_1$ towards the domain of $J(m)$ and $\rho_2 : v \to b_2$ towards the codomain of $J(m)$ must commute, i.e., $J(m) \circ \rho_1 = \rho_2$.*

Definition 6 (Limit). *In a category $\mathcal{C}$, a **limit** is a cone in which the vertex v respects the universal property, stating that for any object $v' \in Ob(\mathcal{C})$ being a vertex of the same cone with components ρ'_i having v' as domain, there exists a unique morphism $\theta : v' \to v$ for which $\rho'_i = \rho_i \circ \theta$. The components of a limit are called the projections.*

Limits are various, and the difference among them is the structure of the index category I. We rely on limits to define some constraints in data models

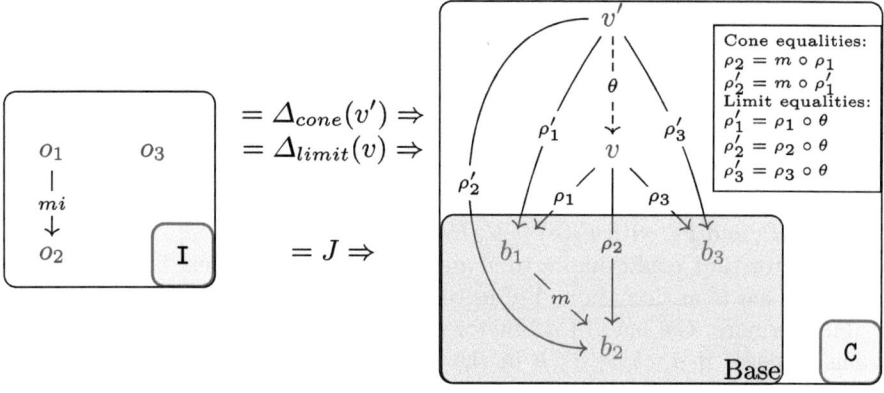

Fig. 1. Representation of a cone and a limit of the index category I, with $\Delta_{cone}(v')$ mapping o_1, o_2 and o_3 to v' and mi to $id_{v'}$, $\Delta_{limit}(v)$ mapping o_1, o_2 and o_3 to v and mi to id_v, and $J(o_1) = b_1$, $J(o_2) = b_2$, $J(o_3) = b_3$ and $J(mi) = m$

and schemas, such as edges in graphs or foreign keys in the relational model (see Sects. 4.3 and 4.4). In figures, the vertices are in purple.

The most common limits in category theory are named and defined, such as the product and the pullback, on which we will rely to represent some constraints.

A product can be seen as a means to get an object a and an object b from an object $a \times b$. Its index category is composed of two objects a, b and no morphism.

Definition 7 (Product). *In a category $\mathcal{C}$, a **product** of two objects a and $b \in Ob(\mathcal{C})$ is an object $a \times b \in Ob(\mathcal{C})$ with two projection morphisms $\rho_1 : a \times b \to a$ and $\rho_2 : a \times b \to b \in Hom_{\mathcal{C}}$. A product can be easily generalized to n objects (i.e., with an index category containing n objects and no morphism).*

A pullback is a commutative diagram that can be seen as a product with a condition over c (i.e., the objects a and b of the pullback depend on c). Its index category is composed of three objects a, b, c and two morphisms $f : a \to c$ and $g : b \to c$.

Definition 8 (Pullback). *A **pullback** in a category $\mathcal{C}$ is a product in which the two objects a and $b \in Ob(\mathcal{C})$ must each have a morphism $f : a \to c$ and $g : b \to c \in Hom_{\mathcal{C}}$ with the same codomain $c \in Ob(\mathcal{C})$. A pullback is composed of an object $a \times_c b \in Ob(\mathcal{C})$ with two projection morphisms $\rho_1 : a \times_c b \to a$ and $\rho_2 : a \times_c b \to b \in Hom_{\mathcal{C}}$, and is commutative, i.e., $f \circ \rho_1 = g \circ \rho_2$.*

3.3 Preservation of Structure

On top of defining mapping among abstraction levels, functors also preserve structure of the source category into the destination category. This structure preservation concerns the domain and codomain of morphisms, the commutative diagrams, and some properties of morphisms linked to identity morphisms.

First, the mapping of a morphism is valid only when the domain and the codomain are the same objects in the source and the destination categories when the functor is applied on them. For example, if we want to map a source category C_1 with two objects $a, b \in Ob(C_1)$ and a morphism $f : a \rightarrow b \in Hom_{C_1}(a, b)$ to a destination category C_2 with two objects $a', b' \in Ob(C_2)$ and a morphism $f' : a' \rightarrow b' \in Hom_{C_2}(a', b')$, there would be three valid functors to do so: F_1, with $F_1(a) = a'$, $F_1(b) = b'$ and $F_1(f) = f'$; F_2, with $F_2(a) = a'$, $F_2(b) = a'$ and $F_2(f) = id_{a'}$; and F_3, with $F_3(a) = b'$, $F_3(b) = b'$ and $F_3(f) = id_{b'}$. However, there is no functor that could map a to b' and b to a', because there is no morphism in C_2 that has b' as domain and a' as codomain.

Furthermore, the laws of a functor also preserve the commutativity of diagrams. Indeed, if $a \circ b = c \circ d$ in the source category, as the preservation of composition law states that $F(h \circ g) = F(h) \circ F(g)$, it means that in the destination category the equalities $F(a) \circ F(b) = F(a \circ b) = F(c \circ d) = F(c) \circ F(d)$ must hold.

Functors also preserve identities with the law stating that $F(id_x) = id_{F(x)}$. For morphisms such as isomorphisms, as $f^{-1} \circ f = id_x$ and $f \circ f^{-1} = id_y$, it means that $F(id_x) = F(f^{-1} \circ f)$ and $F(id_y) = F(f \circ f^{-1})$, and thus that the isomorphism is preserved in the destination category.

The structure preserving property of functors ensures that a data schema only applies structures and constraints available in its data model (see Sect. 5). It also allows to study the migration of a data schema towards a different data model and to deduce the impacts of the migration (see Sect. 6).

4   Categorical Data Models

We propose to rely on category theory to provide a formal framework suited to multi-model data. In this section, we show how categories can represent the structure of a model, and how specific categorical constructs presented in Sect. 3 can be used to embed constraints of models into their category. We also propose categories for the JSON, property graph [6] and relational [13] models. The formalism developed in the section will then be used in Sect. 5 to define categorical data schemas, that are categories linked to their model category with a functor ensuring that the schema complies to its model, and in Sect. 6 to demonstrate how to define a migration with a functor and how to assess its impacts on a schema.

4.1   Representation of Data Models

To represent a data model with a category, we use an object for each element that structures the model (e.g., table, data type, document), and a morphism for a relationship (e.g., attribute that link an element to its data type, link between elements of the structure).

Constraints can be embedded thanks to the mechanisms presented in Sect. 3. Identifiers are represented with an isomorphism, join dependencies are represented with morphism equalities (contained in commutative diagrams), and limits allow to represent various constraints based on a pattern of objects and morphisms. For example, a product can represent an association of elements (such as an edge being an association of two vertices), and a pullback can represent an association with a condition (such as a foreign key linking two attributes over the same domain). To ensure the universal property of the limits, the vertex object representing the constraint is always included in the base itself, with its associated projection being its identity morphism.

Table 1. Summary of category theory notions used to represent data models and schemas

Element	Category theory notion
Basic elements	
Element of structure	Object
Relationship	Morphism
Constraints	
Identifier	Isomorphism (equivalence)
Join dependency	Morphism equality
Advanced constraint	Limit
Specific constraints	
Association of elements of structure	Product
Association of elements of structures with condition	Pullback

The main notions used to represent a data model as a category are summarized in Table 1. We will use the JSON model to demonstrate the applicability of the proposed formalism on a simple data model (i.e., without constraint), as well as the property graph and the relational models to embed constraints of different complexity into the category of the model.

4.2 JSON Model

The category of the JSON model is given in Fig. 2. The main element of this model is a *document*, which can be nested with the morphism *object* : *document* → *document*. A document can contain *typed_attributes*, with the morphism *attribute* : *document* → *typed_attribute*, that can be of type *number*, *string* or *boolean*. An attribute is mapped to its type with the morphisms

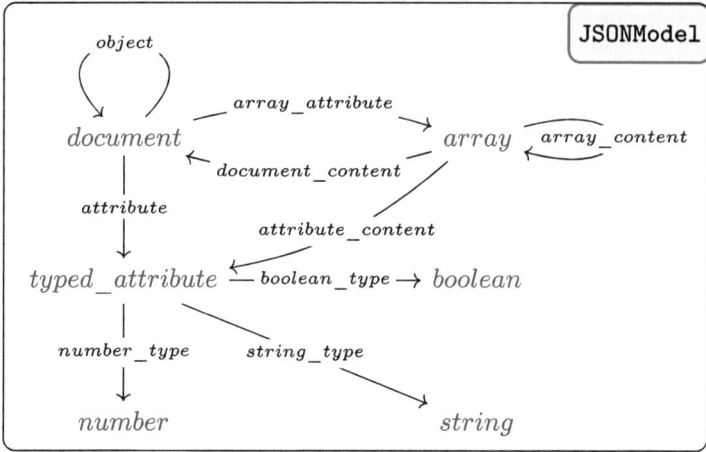

Fig. 2. Category of the JSON model

$number\_type : typed\_attribute \rightarrow number$, $string\_type : typed\_attribute \rightarrow$ $string$ and $boolean\_type : typed\_attribute \rightarrow boolean$. A document can also contain collections of attributes in an $array$ with $array\_attribute : document \rightarrow$ $array$. An array contains documents with the morphism $content : array \rightarrow$ $document$, attributes with $attribute\_content : array \rightarrow typed\_attribute$, or even nested arrays with $array\_content : array \rightarrow array$.

It is important to note that simple attributes are represented with a composition through the object $typed\_attribute$ rather than with a direct morphism from $document$ towards the type of the attribute to ensure compatibility among models that would have typed and/or untyped attributes, and to allow the generalization of attributes through a unique object. Schemas will mainly map their attributes directly to one of the compositions $number\_type \circ attribute$, $string\_type \circ attribute$ or $boolean\_type \circ attribute$ (see Sect. 5).

4.3 Property Graph Model

In the property graph model, both $edge$ and $vertex$ objects can have labels, with the morphisms $edge\_label : edge \rightarrow label$ and $vertex\_label : vertex \rightarrow label$, and properties, with the morphisms $edge\_property : edge \rightarrow typed\_property$ and $vertex\_property : vertex \rightarrow typed\_property$. A property is linked to its type[1] similarly than for the category of the JSON model. Figure 3 shows the category of the property graph model.

An interesting thing to note in the property graph model is the relationship between the objects $edge$ and $vertex$, as an edge is an association of two vertices. The two morphisms $out : edge \rightarrow vertex$ and $in : edge \rightarrow vertex$ indicate which are the source and the destination vertices of an edge. The association constraint

---

[1] Only major types have been represented here.

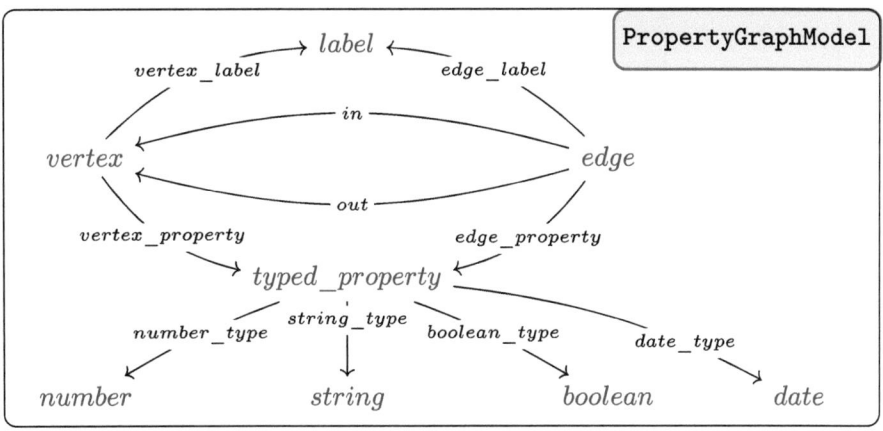

Fig. 3. Category of the property graph model

can be represented with a product having *edge* as the vertex linked to the objects *edge* and *vertex* with the projections id_{edge}, *in* and *out*.

4.4 Relational Model

Compared to the property graph model, the relational model has two main constraints, the primary keys and the foreign keys. It possesses different kinds of attribute, that each needs a specific representation in the category: 1) a simple attribute; 2) an attribute part of a primary key; 3) an attribute part of a foreign key; and 4) an attribute part of a primary key and of a foreign key. Figure 4a shows the category of the relational data model.

Regardless the kind of the attribute, an individual attribute has a domain being the set of values taken by all the tuples of this table for this attribute, represented with the object Dom_{att}. It is mapped to the type of the attribute with a morphism $date\_type : Dom_{att} \rightarrow date$, $string\_type : Dom_{att} \rightarrow string$, $boolean\_type : Dom_{att} \rightarrow boolean$ or $number\_type : Dom_{att} \rightarrow number$. A simple attribute is represented with a morphism $attribute : table \rightarrow Dom_{att}$.

A primary key ensures that all tuples have a different value for a given combination of attributes. Thus, its value alone can be used as an identifier for the whole tuple. We use the isomorphism $pk : table \xleftrightarrow{} Dom_{pk}$ to represent this constraint. For each attribute of a primary key constraint, Dom_{pk} is mapped to the domain of the attribute with the morphism $pk\_att : Dom_{pk} \rightarrow Dom_{att}$.

A foreign key is a strong constraint: it links two tables through an attribute (composed or not), and ensures that this attribute in the referencing table has the same value as a primary key of the referenced table, thus forming a join dependency. A pullback can represent this type of constraint, extracted in Fig. 4b: a foreign key is linked to its referencing table through the morphism *from*, and to the domain of a primary key of its referenced table through the morphism

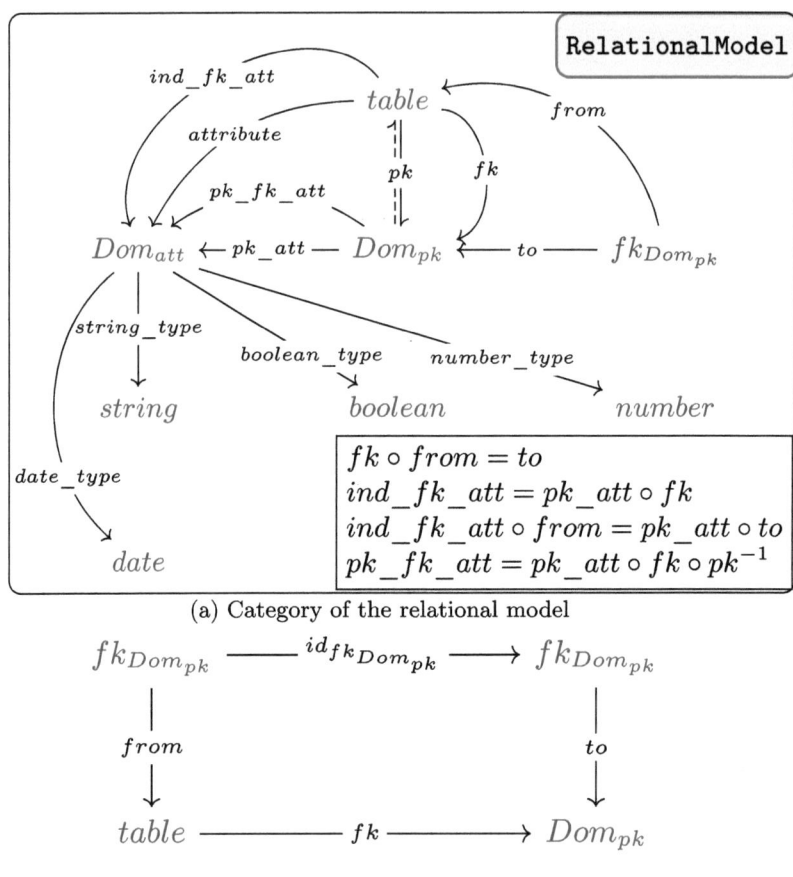

(a) Category of the relational model

$$fk_{Dom_{pk}} \xrightarrow{\quad id_{fk_{Dom_{pk}}} \quad} fk_{Dom_{pk}}$$

$$\downarrow from \qquad\qquad\qquad \downarrow to$$

$$table \xrightarrow{\qquad fk \qquad} Dom_{pk}$$

(b) The pullback representing a foreign key

Fig. 4. Category of the relational model and its foreign key pullback

to. The object Dom_{pk} is the condition of the pullback, it states that the referenced attribute must be a primary key. The fk morphism is the constraint name of the foreign key in the referencing table. The commutativity of the pullback ensures that the join dependency is valid through the equality $fk \circ from = to$. For each attribute of the foreign key, the morphism ind_fk_att is used to map the referencing table to the attribute of the primary key of the referenced table. To ensure the completeness of the join dependency, each attribute are part of two additional commutative diagrams ($ind\_fk\_att = pk\_att \circ fk$ and $ind\_fk\_att \circ from = pk\_att \circ to$) and an associated pullback from $fk_{Dom_{pk}}$ between *table* and $fk_{Dom_{pk}}$ over Dom_{att}.

Finally, some tables might define a primary key with one or more attributes being a foreign key attribute. In which case, the domain of the attribute of the foreign key must be reachable from the domain of the primary key. The morphism $pk\_fk\_att : Dom_{pk} \rightarrow Dom_{att}$ serves this purpose, and the equality $pk\_fk\_att = pk\_att \circ fk \circ pk^{-1}$ ensures that the attribute in the referencing table is indeed equal to the corresponding one in the referenced table.

4.5 Extensibility

The proposed representation of data models can be easily extended. First, new data models can be added by defining a category and using the notions defined in Table 1 to represent constraints and specificities. Second, the formalization can be used to specify particular uses of rules of a model. This extension capability will prove useful to enhance the potential of template migrations between models, for example by defining a template between the relational model towards the property graph model allowing edge properties (see Sect. 6). It can also be used to define particular systems relying on a model but with variations in its application, as for example Oracle DBMS before 23c relies on the relational model but does not propose the boolean type contrary to our `RelationalModel` category.

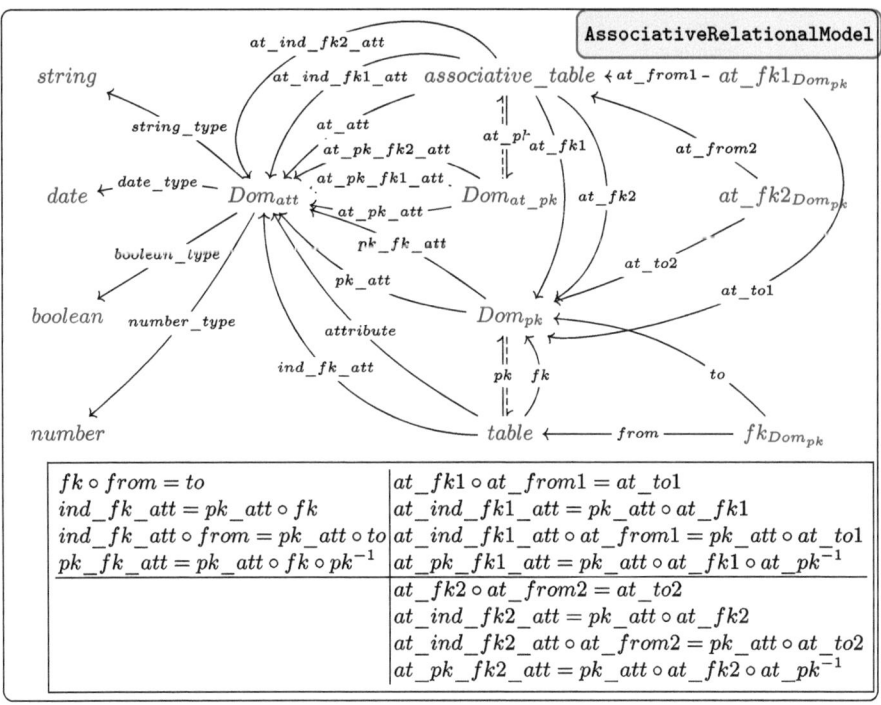

Fig. 5. Category of the relational model with an associative table

To demonstrate how to formalize a specific use of the rules of a data model, we define the category `AssociativeRelationalModel` (Fig. 5).

It differs from the main relational model by defining an *associative_table* that links two tables through foreign keys. As *associative_table*s are still tables, they possess the same capabilities and enforce the same constraints. Once a specific model is defined, a functor must be specified to map it to its main data model. Thanks to the structure preserving capabilities of functors, it ensures that the specific model is compatible with its main data model. The functor mapping the `AssociativeRelationalModel` to the `RelationalModel` category is given in Table 2. It maps the associative table and its associated constructions into the structure of a table.

Table 2. Functor mapping the relational model with an associative table to the relational model

Functor `AssociativeRelationalModel` → `RelationalModel`	
`AssociativeRelationalModel`	`RelationalModel`
Objects	
table, associative_table	*table*
Dom_{pk}, Dom_{at_pk}	Dom_{pk}
Dom_{att}	Dom_{att}
date	*date*
number	*number*
boolean	*boolean*
string	*string*
$fk_{Dom_{pk}}$, $at\_fk1_{Dom_{pk}}$, $at\_fk2_{Dom_{pk}}$	$fk_{Dom_{pk}}$
Morphisms	
pk, at_pk	*pk*
pk_att, at_pk_att	*pk_att*
attribute, at_att	*attribute*
date_type	*date_type*
numer_type	*number_type*
boolean_type	*boolean_type*
string_type	*string_type*
from, at_from1, at_from2	*from*
to, at_to1, at_to2	*to*
fk, at_fk1, at_fk2	*fk*
ind_fk_att, at_ind_fk1_att, at_ind_fk2_att	*ind_fk_att*
pk_fk_att, at_pk_fk1_att, at_pk_fk2_att	*pk_fk_att*

5   Categorical Data Schemas

A schema can apply constraints of its model, and must follow its construction rules. To do so, a schema is formalized both with a category giving the details of the organization of data, and with a functor mapping its category to the category of its corresponding model. The functor defines to which element of a model corresponds each element of a schema, and ensures that the constraints are correctly applied. We will use three different schemas of different complexity to illustrate our proposition: one following the JSON model, one following the property graph model and one following the relational model with an associative table. The functors mapping these three schemas to their respective model are given in Table 3.

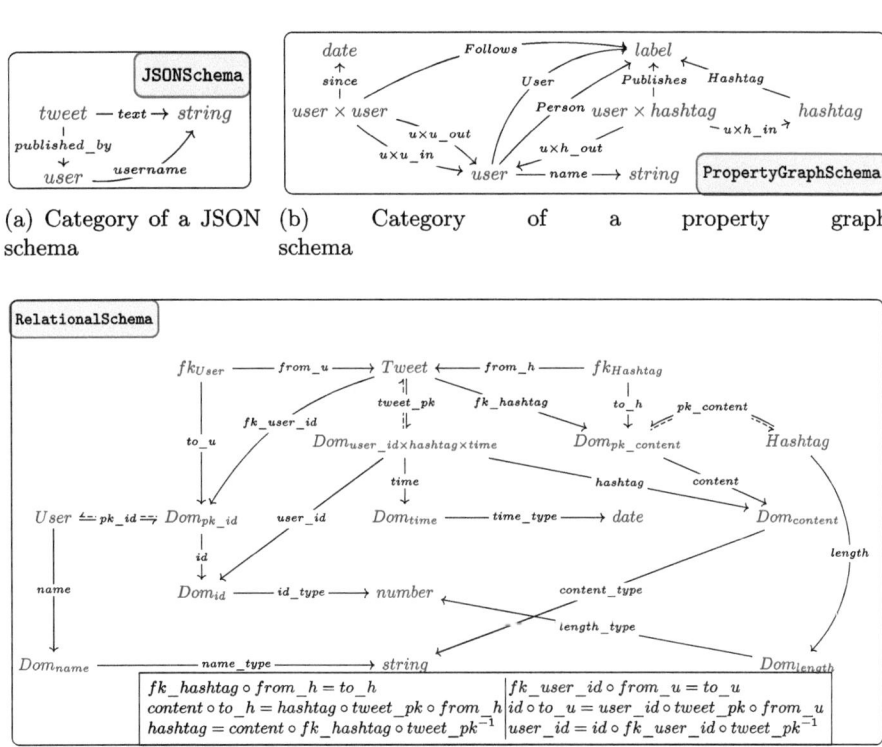

(a) Category of a JSON schema

(b) Category of a property graph schema

(c) Category of a relational schema with an associative table

Fig. 6. Categories of the schemas

Figure 6a shows the category of a simple JSON schema. It contains a document *tweet*, that has an attribute *published_by* representing a nested *user* document. A tweet has a *text*, and a user has a *username*, both attributes of type *string*. There is no specific constraint in this schema.

Figure 6b shows the category of a property graph schema. It contains *user* vertices that have at the same time a label *User* and a label *Person*, and *hashtag* vertices that have a label *Hashtag*. It also has edges with a label *Follows* between two *user* vertices and edges with label *Publishes* between a *user* and a *hashtag*. A vertex *user* has an attribute *name* of type *string*, and an edge *user × user* has an attribute *since* of type *date*. The products of the edge constraints are supported with the objects *u × u_constraint* and *u × h_constraint* as vertices of the products.

Table 3. Functors mapping the schemas to their model

Functor JSONSchema → JSONModel

JSONSchema	JSONModel
Objects	
tweet, user	document
string	string
Morphisms	
published_by	object
text, username	string_type ∘ attribute

Functor PropertyGraphSchema → PropertyGraphModel

PropertyGraphSchema	PropertyGraphModel
Objects	
user × user, user × hashtag	edge
user, hashtag	vertex
string	string
date	date
label	label
Morphisms	
User, Person, Hashtag	vertex_label
Follows, Publishes	edge_label
u × u_in, u × h_in	in
u × u_out, u × h_out	out
name	string_type ∘ vertex_property
since	date_type ∘ edge_property

Functors RelationalSchema → AssociativeRelationalModel **and** RelationalSchema → RelationalModel

RelationalSchema	AssociativeRelationalModel	RelationalModel
Objects		
Tweet	associative_table	table
User, Hashtag	table	table
$Dom_{user\_id \times hashtag \times time}$	Dom_{at_pk}	Dom_{pk}
Dom_{pk_id}, $Dom_{pk\_content}$	Dom_{pk}	Dom_{pk}
Dom_{time}	Dom_{at_att}	Dom_{att}
Dom_{id}, $Dom_{content}$, Dom_{name}, Dom_{length}	Dom_{att}	Dom_{att}
string	string	string
number	number	number
date	date	date
fk_{User}	$at\_fk1_{Dom_{pk}}$	$fk_{Dom_{pk}}$
$fk_{Hashtag}$	$at\_fk2_{Dom_{pk}}$	$fk_{Dom_{pk}}$
Morphisms		
tweet_pk	at_pk	pk
time	at_pk_att	pk_att
user_id	at_pk_fk1_att	pk_fk_att
hashtag	at_pk_fk2_att	pk_fk_att
pk_id, pk_content	pk	pk
id, content	pk_att	pk_att
fk_user_id	at_fk1	fk
fk_hashtag	at_fk2	fk
name_type, content_type	string_type	string_type
id_type, length_type	number_type	number_type
time_type	date_type	date_type
name, length	attribute	attribute
from_u	at_from1	from
from_h	at_from2	from
to_u	at_to1	to
to_h	at_to2	to

Figure 6c shows the category of a relational schema (the corresponding SQL script is given in listing 1.1), with three tables: *User*, *Hashtag* and *Tweet*. *Tweet* is an associative table between the *User* and *Hashtag* tables, and is linked to them with the foreign keys fk_{User} and $fk_{Hashtag}$. *Tweet* has a composed primary key that includes attributes of foreign keys, with the attribute *user_id* being the attribute of fk_{User}, *hashtag* being the attribute of $fk_{Hashtag}$, and *time* being an attribute of type *date*.

Listing 1.1. SQL script of the relational schema

```
CREATE TABLE User (
    id INTEGER,
    name VARCHAR(100),
    CONSTRAINT pk_id PRIMARY KEY(id)
)

CREATE TABLE Hashtag (
    content VARCHAR(100),
    length INTEGER,
    CONSTRAINT pk_content PRIMARY KEY(content)
)

CREATE TABLE Tweet (
    user_id INTEGER,
    hashtag VARCHAR(100),
    time DATE,
    CONSTRAINT fk_user_id FOREIGN KEY (user_id)
        REFERENCES User(id),
    CONSTRAINT fk_hashtag FOREIGN KEY (hashtag)
        REFERENCES Hashtag(content),
    CONSTRAINT tweet_pk
        PRIMARY KEY(user_id, hashtag, time)
)
```

These examples show how the formalization can be adapted to specific schemas. Using a functor to ensure that a schema complies to its model has several benefits. As shown by the JSON schema, a schema does not have to contain all the different elements of its model. Indeed, the definition of the functor ensures that all objects and morphisms of the source category are transformed, but does not force all the objects and morphisms of the destination category to be mapped. A schema can also have multiple elements mapped to a same element of its model, as it is the case for the different edges and vertices of the property graph schema. If a schema follows a specific pattern, such as the relational schema with its associative table, it can be mapped to the general data model (here, the `RelationalModel` category) and to the specific data model (here, the `AssociativeRelationalModel` category).

Furthermore, as stated in Sect. 3.3, a functor preserves the structure of the source category into the destination category. From a data point of view, it ensures that a schema complies to its model, because the structure of the schema category must be valid in the model category to define the functor. It also allows to not use some rules or constraints in some schemas, as the structure preserving capabilities of the functor is only unidirectional.

It is important to note that, even if we do not use an instance representation in our formalization, other works such as [38,44] that have an instance category can be linked to our proposition without modification. Just as for a schema, an instance could be defined in a category, and linked to its schema with a functor.

6   Categorical Data Migrations

Data migrations in multi-model systems are various. Thus, the formal framework must be flexible enough to embrace this variety. To do so, we propose to define template migrations and custom migrations. For both kinds, the migration is represented with a functor from the schema towards the destination model. This functor is used to assess the impact of a migration on a schema, namely if constraints of the schema are preserved and if the migration induces the creation of constraints to comply to the destination model. Thanks to the structure preserving characteristics of functors, their validity will help to perform this assessment. We detail this mechanism and demonstrate its interest on example migrations applied on schemas of the previous section.

6.1   Defining a Migration

A migration is defined by a functor from the source schema towards the destination model. Just as the functor from a schema towards its model, it allows to define into which element of a model each element of the schema will be transformed.

To cope with the variety of migrations in multi-model systems, our proposition allows to define two kinds of migration: 1) through a template migration that is defined between two models; and 2) through a custom migration that is directly defined between the schema of the data to migrate and the destination model. In the first case, all schema elements belonging to a same source model element will be transformed into the same element of the destination model. In the second case, each schema element can be transformed into a different destination model element regardless of the transformation of the other schema elements that belong to the same source model element.

Both kinds of migration are defined with a functor ($F : \mathrm{S} \to \mathrm{DM}$) from the source schema (S) towards the destination model (DM). A template migration ($TF : \mathrm{SM} \to \mathrm{DM}$) is defined from a source model (SM) towards a destination model. When a migration follows a template, F is automatically computed from the composition of the functor mapping the schema to its source model ($MF : \mathrm{S} \to \mathrm{SM}$) and TF. Thus, $\forall x \in Ob(\mathrm{S}), F(x) = TF(MF(x))$, and $\forall f \in Hom_\mathrm{S}, F(f) = TF(MF(f))$. For a custom migration, F is directly defined and does not depend on a composition.

Table 4. Template migration functors from the relational model with and without associative table towards the property graph model

Functors `RelationalModel` → `PropertyGraphModel` **and**
`AssociativeRelationalModel` → `PropertyGraphModel`

RelationalModel	AssociativeRelationalModel	PropertyGraphModel
Objects		
$fk_{Dom_{pk}}$	$fk_{Dom_{pk}}$, $associative\_table$, Dom_{at_pk}, $at\_fk1_{Dom_{pk}}$, $at\_fk2_{Dom_{pk}}$	$edge$
$table$, Dom_{pk}	$table$, Dom_{pk}	$vertex$
		$label$
Dom_{att}	Dom_{att}	$typed\_property$
$string$	$string$	$string$
$number$	$number$	$number$
$date$	$date$	$date$
$boolean$	$boolean$	$boolean$
Morphisms		
	at_pk, at_from1, at_from2	id_{edge}
	at_att, at_pk_att, $at\_ind\_fk1\_att$, $at\_ind\_fk2\_att$, $at\_pk\_fk1\_att$, $at\_pk\_fk2\_att$	$edge\_property$
		$edge\_label$
pk, fk	pk, fk	id_{vertex}
$attribute$, pk_att, ind_fk_att, pk_fk_att	$attribute$, pk_att, ind_fk_att, pk_fk_att	$vertex\_property$
		$vertex\_label$
$from$	$from$, at_to1, at_fk1	out
to	to, at_to2, at_fk2	in
$string\_type$	$string\_type$	$string\_type$
$number\_type$	$number\_type$	$number\_type$
$date\_type$	$date\_type$	$date\_type$
$boolean\_type$	$boolean\_type$	$boolean\_type$

Table 4 shows two examples of template migrations: one from the relational model and one from the relational model with associative table, both towards the property graph model. In the template migration from the relational model, tables are transformed into vertices, and foreign keys into edges. However, there are no morphism that would allow to define edge properties while ensuring the validity of the functor of the template. Nevertheless, when defining a template migration from the relational model with associative table, we can define a way to obtain edge properties. To do so, the associative table is transformed into an edge, and its attributes (other than primary keys) can be transformed into edge properties. The tables that are not associative are transformed into vertices, and their attributes are transformed into vertex properties. This demonstrates the interest of defining specific uses of data models, as it allows the definition of common migrations between models when the schema follows specific patterns.

Table 5. Migration functors from the schema towards the destination model

Functor JSONSchema → RelationalModel

JSONSchema	RelationalModel
Objects	
tweet, user	*table*
string	*string*
Morphisms	
published_by	$pk^{-1} \circ fk$
text, username	*string_type ∘ attribute*

Functors RelationalSchema → PropertyGraphModel *via* AssociativeRelationalModel **and** *via* RelationalModel

RelationalSchema	PropertyGraphModel *via* AssociativeRelationalModel	PropertyGraphModel *via* RelationalModel
Objects		
Tweet, $Dom_{user\_id \times hashtag \times time}$	*edge*	*vertex*
User, Hashtag, Dom_{pk_id}, $Dom_{pk\_content}$	*vertex*	
Dom_{id}, $Dom_{content}$, Dom_{name}, Dom_{length}, Dom_{time}	*typed_property*	
string	*string*	
number	*number*	
date	*date*	
fk_{User}, $fk_{Hashtag}$	*edge*	
Morphisms		
tweet_pk	id_{edge}	id_{vertex}
time, user_id, hashtag	*edge_property*	*vertex_property*
pk_id, pk_content	id_{vertex}	
id, content, name, length	*vertex_property*	
fk_user_id	*out*	id_{vertex}
fk_hashtag	*in*	id_{vertex}
from_u, from_h	id_{edge}	*out*
to_u	*out*	*in*
to_h	*in*	
name_type, content_type	*string_type*	
id_type, length_type	*number_type*	
time_type	*date_type*	

Table 5 gives three example of functor migrations, that we will use in the following subsection to demonstrate how our theoretical framework can assess the impacts of migrations. In the first example, a custom migration is defined to migrate the JSONSchema of Sect. 5 towards the relational model. The objects *tweet* and *user* become two tables, linked through an attribute foreign key *published_by*, and the attributes *text* and *username* become simple attributes of type *string*. The second and third examples apply a template migration on the `RelationalSchema`. The first template is the functor $TF1$: `RelationalModel` $\rightarrow$ `PropertyGraphModel`, and the second template is $TF2$: `AssociativeRelationalModel` $\rightarrow$ `PropertyGraphModel`. To find the migration functor $F1$, we compose the functor $MF1$: `RelationalSchema` $\rightarrow$ `RelationalModel` with $TF1$, and to find the migration functor $F2$, we compose $MF2$: `RelationalSchema` $\rightarrow$ `AssociativeRelationalModel` with $TF2$, as stated at the beginning of the subsection. The difference in the choice of the template is that the $Tweet$ table is either transformed into a *vertex* having links toward vertices $User$ and $Hashtag$, or into an *edge* that links a vertex $User$ to a vertex $Hashtag$.

6.2   Assessing the Impacts of a Migration

We seek to assess two kinds of impacts of a migration on a schema: 1) the loss of constraints of the schema; and 2) the need to create constraints to comply to the destination model.

To verify the preservation of constraints, for each constraint in the schema, the functor of the migration must preserve the entirety of the representation of the constraint. This verification depends on the properties of the category needed to represent the constraint. For commutative diagrams, $F(h \circ g) = F(h) \circ F(g)$ must be verified, i.e., the diagram in the destination model must still commute. For specific morphisms, $F(m)$ must preserve the properties of m. For limits, $\Delta(F(v)) : \text{I} \rightarrow \text{DM}$ and $F \circ J : \text{I} \rightarrow \text{DM}$ must be valid functors for the limit defined in the index category I, i.e., the limit must still exist in the destination model. If the functor does not preserve the representation of the constraint, it means that the destination model does not support this constraint, and that it will be lost during the migration.

Table 6. Objects and morphisms that trigger the verification of the associated constraint for each model

Model	Constraint	Set of objects and morphisms
Property graph	Edge (product)	$\{edge, in, out\}$
Relational	Primary key (isomorphism)	$\{pk, pk^{-1}, Dom_{pk}\}$
	Foreign key (pullback)	$\{fk_{Dom_{pk}}, from, to, fk,$ $ind\_fk\_att, pk\_fk\_att\}$

To verify if a constraint needs to be created, the process is a bit more complex. First, for each constraint of a model, we define the set of objects and morphisms that will be considered as key elements of the constraint. The verification of the need to create the constraint will be performed only if an element of the set is mapped by the functor of the migration. For example, the object Dom_{pk} of the relational model is part of the foreign key constraint, but should not trigger the verification of the need to create a foreign key constraint if it is the only element of the constraint that is mapped. Table 6 gives the sets for the constraints defined for the models of Sect. 4. This step is done only once for each model, regardless of the migration to perform.

Then, if the functor of the migration maps objects and/or morphisms towards an element of a set, the verification of the associated constraint is triggered. For commutative diagrams, it means that if $F(f) = F(g)$ for $F(f), F(g) \in Hom_{DM}$ and $f, g \in Hom_S$, then the equality $f = g$ must hold in the schema category. For specific morphisms, $m \in Hom_S$ must have the same properties as $F(m) \in Hom_{DM}$. For limits from index category I with functor $\Delta(F(v)) : I \to DM$ and $J : I \to DM$, there must be two valid functors $\Delta(v) : I \to S$ and $J' : I \to S$ for which the equalities $J(o) = F(J'(o))$ and $J(m) = F(J'(m))$ hold for each object $o \in Ob(I)$ and morphism $m \in Hom_I$, i.e., the limit must already exists in the schema category. If all the elements of the constraint are correctly mapped and all the conditions of the validity of the constraint are also valid in the schema, the constraint does not need to be created. Otherwise, the constraint needs to be created to guarantee the feasibility of the migration.

We demonstrate this mechanism on the migrations defined in the previous subsection. For the migration represented by the functor JSONSchema $\to$ RelationalModel, there is no preservation of constraint to verify as the schema does not have constraints. However, the migration triggers the verification for the creation of some constraints of the destination model. The morphism pk^{-1} of the relational model is mapped by the morphism *published_by* of the schema, but there is no inverse morphism that maps pk to validate the isomorphism. The morphism fk of the relational model is also mapped by the morphism *published_by*, but the remaining of the pullback of the foreign key constraint is not mapped. Thus, to migrate the JSON schema as specified by the functor, a primary key and a foreign key constraints need to be created, as they are not currently supported by the schema.

For the RelationalSchema $\to$ PropertyGraphModel *via* RelationalModel migration, the isomorphisms of all the primary keys are preserved, but are transformed into identity morphisms. This transformation can be interpreted as the preservation of the constraint, but as an internal mechanism[2]. However, the pullback of the foreign key is not preserved, thus, when considering this schema into the property graph model, it cannot be guaranteed that the attribute *id* for a *user* vertex is equal to the attribute *user_id* of a *tweet* vertex when they are both linked through an edge, resulting in a constraint loss. Regarding the creation of constraint, the product of the *edge* is already supported in the schema thanks to its pullback, thus there is no need to create it.

Finally, the result for the RelationalSchema $\to$ PropertyGraphModel *via* AssociativeRelationalModel migration is almost identical to the previous one, except for the preservation of the pullback of the foreign key. Indeed, for the main

---

[2] For this migration, it means that there is no mean to have the equivalent of a primary key in the property graph model, but that each element can still be internally uniquely identified, as for example with a rowid.

foreign key constraint, the commutativity of the pullback is preserved, but the commutativity for the individual attributes are not preserved. Thus, the equality of the attributes can still not be guaranteed.

7 Prototype and Experimental Evaluation

An open-source prototype[3] has been developed to demonstrate the applicability of the proposed framework, and to automatically assess the impacts of a migration [20]. We give an overview of this prototype, explain how it can be used to improve existing multi-model systems, and perform an experimental evaluation to measure its execution time.

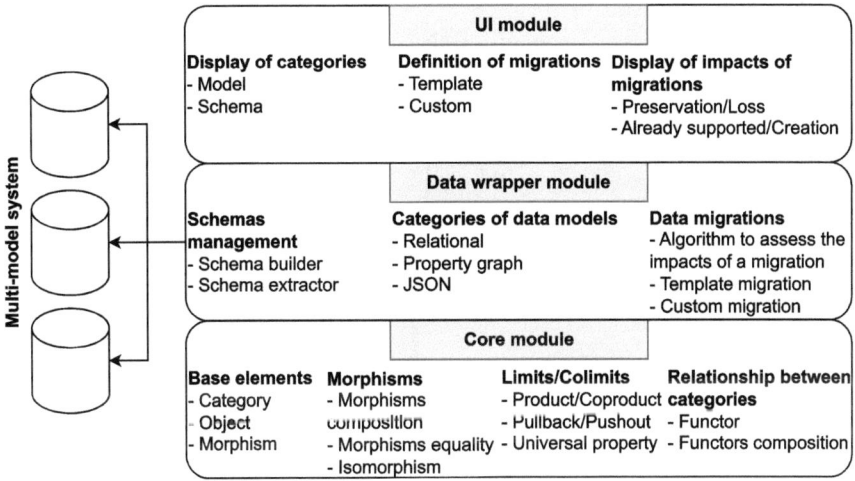

Fig. 7. Main components of the prototype

7.1 Overview of the Implementation

The prototype is developed in Scala (see Fig. 7). It has a core module implementing the different constructs of category theory (objects, morphisms, categories, functors, limits/colimits, etc.), as well as their conditions of construction, such as the functorial laws or the universal property.

This core module is used by a data wrapper module, to: 1) build categories corresponding to data models; and 2) build categories corresponding to given schemas of a model. The available data models are the relational, the property graph and the JSON models. The construction of a schema can be done programatically, or automatically by extracting a schema from a data source connection. The data wrapper module also contains the algorithms used to assess the impacts of a migration on a schema.

---

[3] https://github.com/AnnabelleGillet/CDMiA.

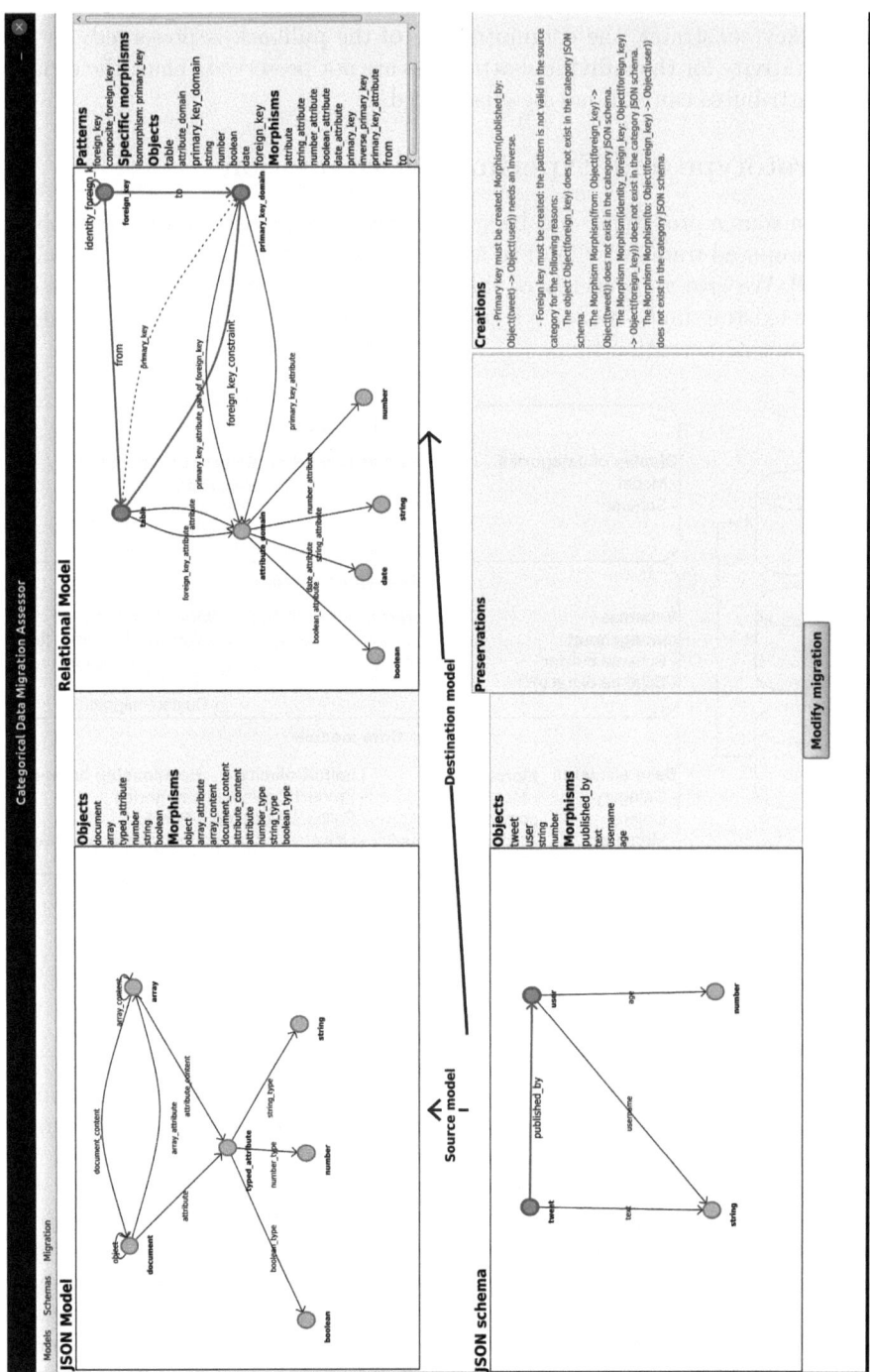

Fig. 8. An example of the output of a migration using our prototype (highlighted elements concern the need to create constraint to support the foreign key between a tweet and a user) (Color figure online)

The prototype has a user interface based on JavaFX and JavaFXSmart-Graph[4], allowing to see categories of data models and schemas, and to define migrations. Users can apply migrations on a schema by choosing a template or by defining a custom migration. The interface displays the impacts of this migration on the selected schema, in order to help data engineers to assess the feasibility of a migration. Figure 8 illustrates the output of the user interface representing the impacts of a migration. The example corresponds to the migration performed on the JSON schema presented in Sects. 5 and 6. The two constraints that must be created to satisfy the migration are displayed in red as they are not currently supported by the schema. When the conditions for the preservation or the creation of a constraint are already fulfilled, they are displayed in green.

7.2   Improvement over Multi-model Systems

Our prototype can be used on top of existing multi-model systems to take into consideration the constraints applied over data schemas and to assess how they will be impacted by a migration.

Migrations can be costly to process, and lead to a high resource and time consumption. Thus, it is essential to prevent beforehand errors provoking the failure of a migration, that can sometimes happen after several hours of execution. As this prototype does not need to perform the migration to assess its impacts, migrations that are doomed to fail can be identified before running them and execution time is not wasted.

Loss of constraints do not necessarily lead to a migration failure (e.g., the loss of a primary key), but they have nonetheless an impact on the informations available in the schema. If a constraint is lost after a migration, and future uses of these data still rely on this constraint, there will be no prevention to ensure that data are correctly manipulated. For example, a primary key prevents to add new data with duplicate values over the primary key attribute, and the loss of a primary key would result in the loss of this prevention. Therefore, warning users when a constraint is lost is essential, even if it does not lead to the migration failure. This prototype identifies this kind of constraint losses, and serves to improve the usability of multi-model systems.

7.3   Experimental Evaluation

We evaluate the execution time of the prototype for loading a schema from a data source and for assessing the impact of a migration on a schema. To do so, we load relational and property graph schemas respectively from a PostgreSQL and a Neo4j database, and we define a migration for each schema by using either template migration `RelationalModel` → `PropertyGraphModel` *via* `RelationalModel` defined in Sect. 6 or template `PropertyGraphModel` →

---

[4] https://github.com/brunomnsilva/JavaFXSmartGraph.

RelationalModel[5]. We used a Dell Precision 5820 (Intel(R) Xeon(R) W-2265 CPU @ 3.50 GHz, 12 cores, 128 GB RAM with Linux kernel 4.19.0-26-amd64) to run the experiments. For reproducibility purpose, the code used to setup and run the experiments is available on Github[6]. We repeat each experiment 10 times and measure the average execution time.

We create several schemas with different parameters, to evaluate their importance on the execution time. In a first experiment, we create only simple elements of schema (such as attributes, documents) without constraint (i.e., tables for the relational schemas and vertices for the property graph schemas). Each simple element has 5 attributes. We vary the number of simple elements from 1 to 1,000 with steps of 100.

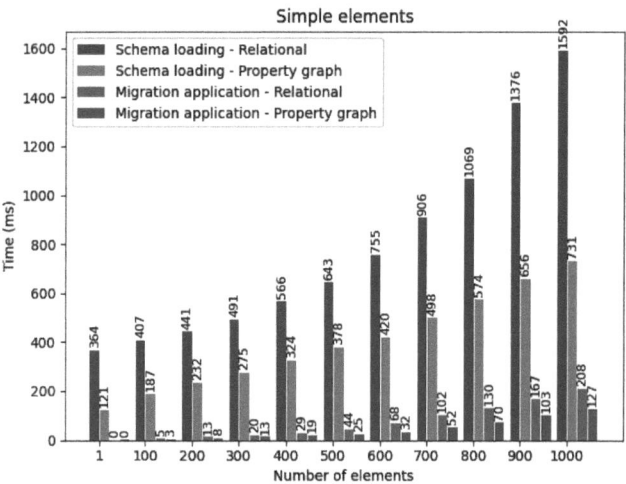

Fig. 9. Execution time for simple elements

Figure 9 shows the result of the first experiment. When there is no constraint preservation or creation to verify, the execution time of the assessment of the impacts of the migration remains low (at most 200ms for 1,000 simple elements) and increases steadily depending on the number of elements in the schema. The schema extraction time also increases depending on the number of elements, but is reasonable (1.5s at most for 1,000 elements).

In a second experiment, we add constraints in the schemas. For each schema, there are 100 simples elements (i.e., 100 tables in the relational schemas and 100 vertices in the property graph schemas). In relational schemas, each table has a primary key. We create from 1 to 10 foreign keys in each table (so from 100

---

[5] Its definition can be found at https://github.com/AnnabelleGillet/CDMiA/blob/main/experiments/benchmark_execution_neo4j.scala. It maps *vertex* to *table*, and *edge* to $fk_{Dom_{pk}}$.

[6] https://github.com/AnnabelleGillet/CDMiA/tree/main/experiments.

to 1,000 foreign keys in schemas), and from 100 to 1,000 edges in the property graph schemas that are distributed among vertices.

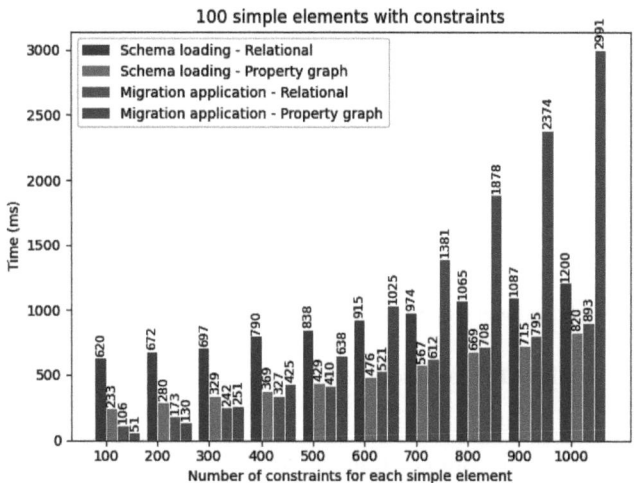

Fig. 10. Execution time for 100 elements with constraints

Figure 10 shows the result of the second experiment. The execution time for the relational schemas increases steadily, but it is higher for the property graph schemas (less than 900ms for 1,000 constraints in the relational schema and less than 3s for 1,000 constraints in the property graph schema). The difference can be explained by the mapping of the migrations. When migrating from a relational schema towards the property graph model, each foreign key (represented by a pullback) is mapped to an edge (represented with a product), whereas when migrating from a property graph schema towards the relational model, each edge is mapped to a foreign key. The pullback is a constraint that is more complex to verify than the product, because the morphism equality must be checked on top on the correctness of the pattern. Thus, the execution time is more impacted by the complexity of the destination category than the complexity of the source category.

In a third experiment, we vary the volume of data with a fixed number of elements in the schemas. We use the same schema as the first experiment, and we vary the number of tuples in the relational schemas and of vertices in the property graph schemas from 1 to 100,000 by multiplying the volume by 10 at each step.

Figure 11 shows the result of the third experiment. The assessment of the impacts of a migration is not impacted at all by the data volume. However, the schema extraction depends on the DBMS: when the DBMS explicitly stores the schema (such as PostgreSQL), the execution time does not increase, but when the DBMS does not store the schema (such as Neo4j), the execution time

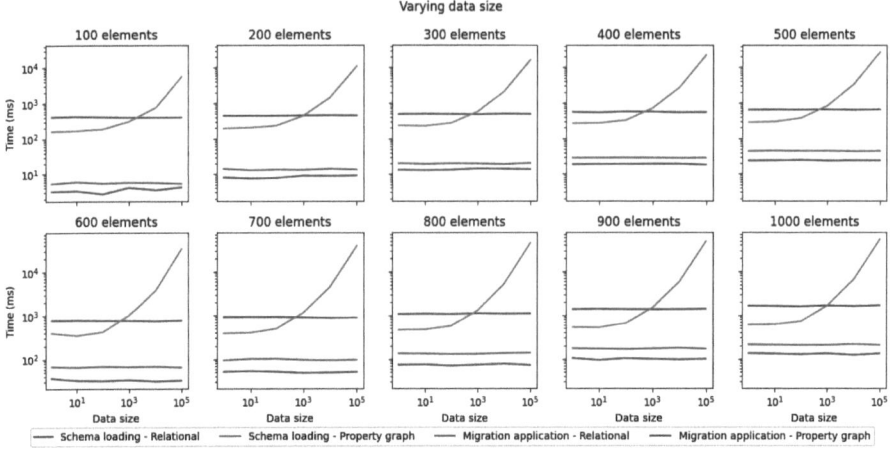

Fig. 11. Execution time for varying data volume

increases with the data volume. Indeed, Neo4j infers the schema by sampling data. It is important to note that the schema extraction step is out of scope of this paper, and we rely on existing works to support this step.

8   Conclusion

We proposed to rely on category theory to represent data models with categories, data schemas with categories mapped to their model through a functor, and data migrations with functors from the schema towards the destination model. This representation **ensures that a schema complies to a model**, and allows to **assess the impacts of a migration**, whether it induces a constraint loss or a need to create a constraint of the destination model. This assessment can be done even if the constraints in the source and in the destination models are not exactly the same, as long as they share the properties that are being verified.

Contrary to most approaches that focus on providing a unified schema regardless of its model, the representation of each model and its constraints is directly integrated into our formalization. It allows to **embrace the variety of multi-model systems without losing information brought by the model itself**. Furthermore, the formalization is easily extensible, and can support the addition of other data models or specific uses of a data model.

As future work, we plan to add querying capabilities to our formalization. It could allow to define operators of models, and to deduce equivalences among operators regardless of their model. Thus, it would open up perspectives for query optimizations, and for improving polyglot capabilities of multi-model systems.

References

1. Abiteboul, S., et al.: Research directions for principles of data management. Dagstuhl Manifestos **7**(1), 1–29 (2018)
2. Alagić, S., Bernstein, P.A.: A model theory for generic schema management. In: Ghelli, G., Grahne, G. (eds.) DBPL 2001. LNCS, vol. 2397, pp. 228–246. Springer, Heidelberg (2002). https://doi.org/10.1007/3-540-46093-4_14
3. Allaire, P., Augat, J., Jose, J., Merrill, D.: Reducing costs and risks for data migrations. Santa Clara, CA, USA **31**, 1–26 (2010)
4. Alotaibi, R., Bursztyn, D., Deutsch, A., Manolescu, I., Zampetakis, S.: Towards scalable hybrid stores: constraint-based rewriting to the rescue. In: Proceedings of the 2019 International Conference on Management of Data, pp. 1660–1677 (2019)
5. Alotaibi, R., Cautis, B., Deutsch, A., Latrache, M., Manolescu, I., Yang, Y.: Estocada: towards scalable polystore systems. Proc. VLDB Endow. **13**(12), 2949–2952 (2020)
6. Angles, R., et al.: PG-schema: schemas for property graphs. Proc. ACM Manag. Data **1**(2), 1–25 (2023)
7. Barret, N., Manolescu, I., Upadhyay, P.: Abstra: toward generic abstractions for data of any model. In: Proceedings of the 31st ACM International Conference on Information and Knowledge Management, pp. 4803–4807 (2022)
8. Bernstein, P.A.: Generic model management: a database infrastructure for schema manipulation. In: Batini, C., Giunchiglia, F., Giorgini, P., Mecella, M. (eds.) CoopIS 2001. LNCS, vol. 2172, pp. 1–6. Springer, Heidelberg (2001). https://doi.org/10.1007/3-540-44751-2_1
9. Bernstein, P.A., Haas, L.M., Jarke, M., Rahm, E., Wiederhold, G.: Panel: is generic metadata management feasible? In: VLDB, pp. 660–662 (2000)
10. Bernstein, P.A., Madhavan, J., Rahm, E.: Generic schema matching, ten years later. VLDB Endow. **4**(11), 695–701 (2011)
11. Bernstein, P.A., Melnik, S.: Model management 2.0: manipulating richer mappings. In: ACM SIGMOD, pp. 1–12 (2007)
12. Bernstein, P.A., Halevy, A.Y., Pottinger, R.A.: A vision for management of complex models. ACM SIGMOD Rec. **29**(4), 55–63 (2000)
13. Codd, E.F.: A relational model of data for large shared data banks. Commun. ACM **13**(6), 377–387 (1970)
14. Conrad, A., Utzmann, P., Klettke, M., Störl, U.: Metamodels to support database migration between heterogeneous data stores. In: Proceedings of the 25th International Conference on Model Driven Engineering Languages and Systems: Companion Proceedings, pp. 546–551 (2022)
15. Damiani, E., Oliboni, B., Quintarelli, E., Tanca, L.: A graph-based meta-model for heterogeneous data management. Knowl. Inf. Syst. **61**, 107–136 (2019)
16. Dasgupta, S., Coakley, K., Gupta, A.: Analytics-driven data ingestion and derivation in the awesome polystore. In: 2016 IEEE International Conference on Big Data (Big Data), pp. 2555–2564. IEEE (2016)
17. Diskin, Z., Kadish, B.: Algebraic graph-oriented= category-theory-based. Manifesto of categorizing database theory. Technical Report, 9406 (1996)
18. Gadepally, V., et al.: The BigDAWG polystore system and architecture. In: 2016 IEEE High Performance Extreme Computing Conference (HPEC), pp. 1–6. IEEE (2016)
19. Gadepally, V., et al.: D4m: bringing associative arrays to database engines. In: 2015 IEEE High Performance Extreme Computing Conference (HPEC), pp. 1–6. IEEE (2015)

20. Gillet, A., Leclercq, É.: CDMiA: revealing impacts of data migrations on schemas in multi-model systems. In: Islam, S., Sturm, A. (eds.) CAiSE 2024. LNCS, vol. 520, pp. 120–128. Springer, Cham (2024). https://doi.org/10.1007/978-3-031-61000-4_14

21. Gillet, A., Leclercq, É., Cullot, N.: Lambda+, the renewal of the lambda architecture: category theory to the rescue. In: La Rosa, M., Sadiq, S., Teniente, E. (eds.) CAiSE 2021. LNCS, vol. 12751, pp. 381–396. Springer, Cham (2021). https://doi.org/10.1007/978-3-030-79382-1_23

22. Hai, R., Koutras, C., Quix, C., Jarke, M.: Data lakes: a survey of functions and systems. IEEE Trans. Knowl. Data Eng. (2023)

23. Halevy, A., Franklin, M., Maier, D.: Principles of dataspace systems. In: Proceedings of the Twenty-Fifth ACM SIGMOD-SIGACT-SIGART Symposium on Principles of Database Systems, pp. 1–9 (2006)

24. Jarke, M., Quix, C.: Federated data integration in data spaces. In: Designing Data Spaces, p. 181 (2022)

25. Johnson, M., Dampney, C.N.: On the value of commutative diagrams in information modelling. In: Nivat, M., Rattray, C., Rus, T., Scollo, G. (eds.) Algebraic Methodology and Software Technology (AMAST'93). Workshops in Computing, pp. 45–58. Springer, London (1994). https://doi.org/10.1007/978-1-4471-3227-1_5

26. Kensche, D., Quix, C., Chatti, M.A., Jarke, M.: GeRoMe: a generic role based metamodel for model management. J. Data Semant. VIII, 82–117. Springer (2007)

27. Kolev, B., Bondiombouy, C., Valduriez, P., Jiménez-Peris, R., Pau, R., Pereira, J.: The CloudMdsQL multistore system. In: Proceedings of the 2016 International Conference on Management of Data, pp. 2113–2116 (2016)

28. Kolev, B., et al.: Parallel polyglot query processing on heterogeneous cloud data stores with LeanXcale. In: 2018 IEEE International Conference on Big Data (Big Data), pp. 1757–1766. IEEE (2018)

29. Koupil, P., Holubová, I.: A unified representation and transformation of multi-model data using category theory. J. Big Data 9(1), 61 (2022)

30. Lenzerini, M.: Direct and reverse rewriting in data interoperability. In: Giorgini, P., Weber, B. (eds.) CAiSE 2019. LNCS, vol. 11483, pp. 3–13. Springer, Cham (2019). https://doi.org/10.1007/978-3-030-21290-2_1

31. Liu, Z.H., Lu, J., Gawlick, D., Helskyaho, H., Pogossiants, G., Wu, Z.: Multi-model database management systems - a look forward. In: Gadepally, V., Mattson, T., Stonebraker, M., Wang, F., Luo, G., Teodoro, G. (eds.) DMAH/Poly -2018. LNCS, vol. 11470, pp. 16–29. Springer, Cham (2019). https://doi.org/10.1007/978-3-030-14177-6_2

32. Lu, J., Holubová, I.: Multi-model databases: a new journey to handle the variety of data. ACM Comput. Surv. (CSUR) 52(3), 1–38 (2019)

33. Manolescu, I.: Understanding and querying data regardless of the data model. In: VLDB Summer School 2023 (2023)

34. Nelson, D., Rossiter, B., Heather, M.A.: The functorial data model-an extension to functional databases. Technical report Series, Department of Computing Science (1994)

35. Schultz, P., Spivak, D.I., Vasilakopoulou, C., Wisnesky, R.: Algebraic databases. Theory Appl. Categories 32(16), 547–619 (2017)

36. She, Z., Ravishankar, S., Duggan, J.: Bigdawg polystore query optimization through semantic equivalences. In: 2016 IEEE High Performance Extreme Computing Conference (HPEC), pp. 1–6. IEEE (2016)

37. Spivak, D.I.: Functorial data migration. Inf. Comput. 217, 31–51 (2012)

38. Spivak, D.I.: Category Theory for the Sciences. MIT Press, Cambridge (2014)
39. Spivak, D.I., Kent, R.E.: Ologs: a categorical framework for knowledge representation. PLoS ONE **7**(1), e24274 (2012)
40. Stonebraker, M., Çetintemel, U.: "One size fits all" an idea whose time has come and gone. In: Making Databases Work: The Pragmatic Wisdom of Michael Stonebraker, pp. 441–462 (2018)
41. Störl, U., Klettke, M.: Darwin: a data platform for noSQL schema evolution management and data migration. In: Workshop Proceedings of the EDBT/ICDT 2022 (2022)
42. Svoboda, M., Čontoš, P., Holubová, I.: Categorical modeling of multi-model data: one model to rule them all. In: Attiogbé, C., Ben Yahia, S. (eds.) MEDI 2021. LNCS, vol. 12732, pp. 190–198. Springer, Cham (2021). https://doi.org/10.1007/978-3-030-78428-7_15
43. Tan, R., Chirkova, R., Gadepally, V., Mattson, T.G.: Enabling query processing across heterogeneous data models: a survey. In: 2017 IEEE International Conference on Big Data (Big Data), pp. 3211–3220. IEEE (2017)
44. Uotila, V., Lu, J.: A formal category theoretical framework for multi-model data transformations. In: Rezig, E.K., et al. (eds.) DMAH Poly 2021. LNCS, vol. 12921, pp. 14–28. Springer, Cham (2021). https://doi.org/10.1007/978-3-030-93663-1_2
45. Uotila, V., Lu, J., Gawlick, D., Liu, Z.H., Das, S., Pogossiants, G.: Multi-model query processing meets category theory and functional programming. In: SEA-Data@ VLDB, pp. 48–49 (2021)
46. Wisnesky, R., Spivak, D.: A Functorial Query Language. Boston Haskell (2014)
47. Yu, X., Gadepally, V., Zdonik, S., Kraska, T., Stonebraker, M.: FastDAWG: improving data migration in the BigDAWG Polystore system. In: Gadepally, V., Mattson, T., Stonebraker, M., Wang, F., Luo, G., Teodoro, G. (eds.) DMAH/Poly -2018. LNCS, vol. 11470, pp. 3–15. Springer, Cham (2019). https://doi.org/10.1007/978-3-030-14177-6_1

Consistently Mapping Differential Privacy Paradigms Between Relational Databases and RDF

Sara Taki[(✉)], Adrien Boiret, Cédric Eichler, and Benjamin Nguyen

Laboratoire d'Informatique Fondamentale d'Orléans, INSA Centre Val de Loire,
Université d'Orléans, PETSCRAFT project-team, Inria, Bourges, France
{sara.taki,adrien.boiret,cedric.eichler,benjamin.nguyen}@insa-cvl.fr

Abstract. The Semantic Web represents an extension of the current web offering a metadata-rich environment based on the Resource Description Format (RDF) which supports advanced querying and inference. However, relational database (RDB) management systems remain the most widespread systems for (Web) data storage. Consequently, the key to populating the Semantic Web is the mapping of RDB to RDF, supported by standardized mechanisms. Confidentiality and privacy represent significant barriers for data owners when considering the translation and subsequent utilization of their data. In order to facilitate acceptance, it is essential to build privacy models that are equivalent, explainable, and usable within both data formats.

Differential Privacy (DP) has emerged to be the flagship of data privacy when sharing or exploiting data. Recent works have proposed DP-models tailored for either multi-relational databases or RDF. This paper leverages this field of work to study how privacy guarantees on RDB with foreign key constraints can be transposed to RDF databases and vice versa.

We use classical RDB and RDF formalisms as well as an established translation tool between the two (RDB2RDF) to compare DP models between the two data representations. We consider a promising DP model for RDB related to cascade deletion and demonstrate that it is sometimes similar to an existing DP graph privacy model, but inconsistently so. Consequently, we tweak this model in the relational world and propose a new model called *restrict deletion*. We show that it is equivalent to an existing DP graph privacy model, facilitating the comprehension, design and implementation of DP mechanisms in the context of the mapping of RDB to RDF. Conversely, we consider a useful DP model with label differentiation capabilities on graphs (QL-outedge), and propose its transposition into an original RDB distance.

Keywords: Differential Privacy · Relational Databases · Knowledge Graph · RDB2RDF mapping

1 Introduction

The Semantic Web represents an extension of the current web standardized by the World Wide Web Consortium. It relies on the Resource Description

© The Author(s), under exclusive license to Springer-Verlag GmbH, DE, part of Springer Nature 2026
A. Hameurlain et al. (Eds.): *Transactions on Large-Scale Data- and Knowledge-Centered Systems LIX*, LNCS 16240, pp. 94–121, 2026.
https://doi.org/10.1007/978-3-662-72449-1_4

Format (RDF) and provides metadata-rich, reusable, and shareable data. RDF is a form of knowledge graph that can be coupled with ontologies such as OWL, thereby enhancing the semantic value of the data and inference capabilities. A key advantage of the Semantic Web is its ability to enrich data with well-defined semantics and to interconnect datasets through the RDF ontology language. A major motivation for this interlinking lies in the identification and integration of heterogeneous related data sources, significantly enhancing the value of using open datasets. As pointed out by reports such as Field *et al.* [1] and Michel *et al.* [2], many domains, such as neuroscience, biology, or social sciences, require combining and analyzing datasets that span multiple scales and representations. Achieving this integration demands explicit semantics that allow diverse databases to be interpreted in a common framework.

Currently, vast volumes of data still reside in relational databases (RDB), and relational database management systems (RDBMS), such as Oracle and PostgreSQL, remain among the most popular systems to manage data[1] Mapping data from relational databases to RDF is a key to populate the Semantic Web. Transforming this massive amount of data into a machine-readable format is likely to facilitate the integration of various data sources and the emergence of new applications and innovative technological solutions. Mapping has been an active field of research during the last two decades [2–4] initiated by the RDB2RDF (Relational Database to Resource Description Format) incubator group[2].

However, the benefits of data integration and sharing also raise new concerns. Data collected (whether stored in RDB or RDF) can contain sensitive information. With the increasing attention on data privacy and the development of privacy regulations (e.g., the General Data Protection Regulation in the European Union), it is becoming increasingly important to protect sensitive information when sharing or allowing the utilization of data. Such concerns are a significant obstacle for RDB holders in accepting the translation and subsequent utilization of their data, as protection models vastly differ between RDF and RDB formats. *It is crucial to construct models that are equivalent and explainable within both formats, easing the comprehension of the guarantees provided in RDF within the familiar context of RDB.*

Differential Privacy (DP) [5] is a classical yardstick to measure privacy protection. Publication mechanisms that satisfy DP provide a form of indistinguishability. That is to say, it is difficult for an entity observing the output of a DP mechanism to determine which of several *adjacent* or *neighboring* databases was used as input. If two neighboring databases differ by the contribution of an individual, an external observer may therefore not know with high confidence whether the data pertaining to a particular individual has been used. Hence, it may not infer anything significant on such data. The concept of adjacency is thus a cornerstone of DP, defining what is protected. In the most simple context, a database is a single, monolithic table of records (or tuples) that holds

---

[1] See: https://db-engines.com/en/ranking.

[2] http://www.w3.org/2001/sw/rdb2rdf/.

private data. In this context, neighboring databases are those that differ by one record, meaning that DP protects (or "hides") the presence or absence of any single record in the database [5]. The intuition here is that each individual participates in, at most, one database record and therefore DP indeed protects the contribution of each individual.

Defining neighborhoods for multi-relational databases, i.e., databases composed of many tables, is challenging for many reasons (see, for instance, [6]). The introduction of several relations usually comes with *constraints*, each constraint arising from the semantics of the database. It is thus no longer possible to define an adjacent database simply by adding or removing a tuple in a table since this may violate the database constraints. In this paper, we consider an important type of constraint, foreign key (FK) constraints, sometimes associated to cardinality constraints.

In graphs, however, there exists several notions of distances where neighbors differ only locally, the most classic being the node distance (where two neighboring graphs differ only by the deletion of a node and its incident edges) and the edge distance (where two neighboring graphs differ only by the deletion of a single edge). We note that those different distances create different neighborhoods, and it turn, lead to DP models that provide different guarantees.

This paper focuses on distances and neighborhoods in the RDB and RDF worlds that are equivalent through standard translation mechanisms to build DP-models that are equivalent in both. More specifically, we focus on the following objectives:

O1. Establish a framework to match or compare DP models between the RDB and RDF formalisms. DP models can be characterized by their distance, which means we want mappings of databases from one formalism to the other with some distance-preserving properties.

O2. Apply this framework to the DP model on RDB centered around cascade deletion [7]. We aim to find a matching RDF DP model, and study potential adjustments.

O3. Apply this framework to the DP model on RDF centered around outedges of specific labels [8]. We aim to find a matching RDP DP model, and characterize mapping choices for the comparison to be meaningful.

The fulfillment of these objectives would offer the benefit of a common ground that actors considering DP in RDB and RDF can share, and leverage it to offer a variety of DP models, actable in both formalisms, that can provide a wide variety of privacy guarantees, depending on a specific application's requirements.

Contributions

To meet these objectives, this paper:

– formalizes the notion of encoding (or mapping) from RDB to RDF, consistent with R2RML mapping, that subsumes standard-compliant W3C recommendations.

– formalizes a generalized notion of cascade deletion in RDB covering both transitive deletion and our proposal.
– studies the translation in RDF of the DP model relying on transitive deletion.
– introduces a meaningful relaxation of this model and demonstrates that it is equivalent to an existing graph privacy model through mapping.
– proposes a RDB model and R2RML encoding choices that make it equivalent to QL-Outedge privacy [8], which provides semantics for the neighborhoods.

The remainder of this paper is structured as follows. The next section details relevant related works on mapping and DP. Section 3 introduces an illustrative example as well as the formalization of the considered databases and mappings. Section 4 proposes an analysis of the translation of transitive deletion in RDF and details our proposed relaxation, demonstrating its equivalence to a well-established DP model in graphs through mapping. Section 5 proposes a novel distance in relational databases which, under adequate encoding, is equivalent to the concept of QL-Outedge privacy. Adequacy of these encodings and their impact on privacy is further illustrated in Sect. 6. Finally, Sect. 7 concludes this paper and discusses future work.

This paper is an extended version of our previous works [9,10]. In [9], we first introduced the problem addressed here, but only in an informal manner and without formal proofs of model equivalency through mappings. A first formalization, focusing on cascade deletion, was later proposed in [10]. Building on these foundations, the present paper extends the formalization of RDB and RDF to the encodings that connect them, and broadens the comparative study of distances to include QL-outedge. This requires particular choices in R2RML mappings in order to admit an equivalent in RDB.

2   Related Work

This section introduces background on RDB to RDF mapping, before introducing DP and its adaptations to RDB and RDF databases. *To the best of our knowledge, this paper is the first at the intersection of these two fields*, focusing on the impact of the RDB to RDF translation on DP-models and the definition of equivalent DP-models in both worlds.

2.1   Mapping RDB to RDF

In September 2012, the RDB2RDF Working Group published two Recommendations: Direct Mapping (DM) [11] and customized mapping (CM) R2RML [12].

The W3C DM recommendation defines simple mapping rules to map relational data to RDF [13]. The RDF generated straightforwardly is based on the structure of the database schema. URIs are automatically generated [14]. Many-to-many relations in relational databases are generally represented as a join table where all its columns are foreign keys (FKs) to other tables (n-ary relations). One missing part from the DM is to represent many-to-many relations as simple triples [15]. When DM is applied, the join table will be translated into a distinct class, which conflicts with the canonical representation of many-to-many relationship in RDF.

CM R2RML [12] is a RDB to RDF mapping language that allows to manually customize the mapping. The expert user has to know the RDB and the domain ontology to express the schema utilizing an existing target ontology. The W3C RDB2RDF Working Group proposed a set of core requirements for R2RML [16], including the exposition of many-to-many join tables as simple triples [15].

Due to the representation of many-to-many join tables as simple RDF triples, we consider mapping mechanisms conform to the R2M2RL specifications.

Extended Mapping Models. RML [17,18] extends the R2RML mapping language to support the mapping of data sources with diverse formats, including data formats like XML, CSV/TSV and JSON. However, it does not tackle the constraints associated with handling various kinds of databases and query languages. xR2RML [19], a mapping language developed as an extension of R2RML and RML. Beyond relational databases, xR2RML also supports the mapping of many non-relational databases to RDF. It is intended to flexibly adjust to diverse data models and query languages. In addition, it can handle data under heterogeneous formats. In the rest of the article, we focus on R2RML as it is the original standard, but adapting our work to cover extensions such as RML and xR2RML could be an interesting perspective.

2.2   Differential Privacy

Differential Privacy (DP) proposes a robust mathematical framework for privacy protection [20]. An algorithm respects DP if observing its output does not permit to determine with strong confidence which of several neighboring dataset was used as input.

Definition 1 (ϵ-differential privacy). *Given $\epsilon > 0$, a function $f : \mathcal{X} \to \mathcal{S}$ and a distance d over $\mathcal{X}$, is ϵ-differentially private if, for any couple of datasets $(D, D') \in \mathcal{D}^2$ such that $d(D, D') = 1$, and for any $S \subseteq \mathcal{S}$:*

$$Pr[f(D) = S] \leq e^{\epsilon} \times Pr[f(D') = S]$$

where probability Pr is over the randomness of f.

Parameter ϵ is also known as the privacy budget, a smaller value indicating stricter privacy requirements. Two datasets at a distance one are said to be neighbors. One classical way of achieving DP for a function (e.g. a query) f is to add an appropriate amount of noise to its results, calibrated by the *global sensitivity* (GS) of f. GS measures the maximal variation of the query result when evaluated upon any two neighboring databases.

Definition 2 (Global Sensitivity (GS) [5]). *For a function $f : \mathcal{D} \to \mathcal{S}$ and a distance d over $\mathcal{X}$ for all datasets $(D, D') \in \mathcal{D}^2$:*

$$df = GS_f = \max_{D,D':d(D,D')=1} \| f(D) - f(D') \|_1$$

where $\|\|_1$ denotes the L1 norm.

GS depends only on *f*, the considered space of databases $\mathcal{X}$, and the distance *d* it is associated with (i.e., that identify neighboring databases). It is independent of the database itself. For queries with low GS, only a small magnitude of noise needs to be added to respect DP. On the other hand, when the GS is high, a substantial amount of noise must be injected to achieve DP, which will impair data utility.

DP for RDF. DP is immediately applicable to any space $\mathcal{X}$ given a proper distance *d* or notion of neighborhood over $\mathcal{X}$. When considering graphs, two models prevail: *node-DP* and *k-edge-DP*. We consider a third, more recent, proposal integrating semantics, *QL-edge-labeled-DP*.

In node-DP, neighboring graphs differ by a single node and all its incident edges, thereby protecting each node along with its incident edges. While node-DP is the strongest of these models, it poses a particular challenge: two neighboring graphs can differ by an arbitrarily large number of edges, which may lead to high variations in outputs across the neighborhood and result in low-utility DP mechanisms.

k-edge-DP [21] is a looser model in which two graphs are adjacent if they differ by up to k edges. Compared to node privacy, edge privacy is limited to protecting *k* relationships. 1-edge-DP is simply called edge-DP and the most commonly employed in the literature.

We note that these two models do not discriminate nodes or edges based on attributes or labels. Such considerations have value in privacy, as information is not uniformly sensitive: for instance, the timestamp of a tweet is usually considered less sensitive than a personal address. Reuben [7] studied the adaptation of DP to edge-labeled directed graphs by defining sets of sensitive labels to which the protection is restricted. Reuben introduced the notion of QL-edge-label neighboring graphs: graphs that differ by a set of outedges of a node with specific labels. The underlying idea behind this definition is that, for example, in RDF graphs, some relations of an entity may be innocuous and some should be considered sensitive, shown by particular labels. Given this neighboring definition, the author presented **QL-edge-labeled-DP**, that considers edges' semantics by only protecting edges of a given set *QL* (i.e., *sensitive* labels). As such, it only protects a predetermined subset of edges. This notion can be transposed to most models. For example, k-typed-edge DP could be defined as the model where two adjacent graphs differ by up to k *sensitive* edges.

Reuben considers the outedges of a node since it denotes the contributions made by that node within the graph. This semantically grabs the idea of the presence of an individual in a graph while not being present in another graph, analogous to how private data is modeled as a tuple in the relational model. Throughout this article, we will use **QL-Outedge DP** as a synonym for **QL-edge-labeled DP** to better reflect the considered model. The related distance for QL-Outedge privacy is presented in [8].

DP for Multi-relational Databases. In the DP literature [5], a database is commonly a single, monolithic table of records (or tuples) that holds private data. Multi-relational databases, i.e., databases composed of many tables, are less popular. However, DP has also been investigated in this setting [22–24].

PINQ [22] and FLEX [23] consider a simple definition of neighboring databases, which does not consider foreign key (FK) constraints. According to their definition, neighboring databases possess the same set of relations and attributes and differ by exactly one tuple in one relation.

The PrivateSQL system [24] introduces a richer notion of neighboring databases that considers constraints in the schema, in particular primary and FK constraints. Under this model, upon deletion of one tuple from one relation, many tuples in other relations have to be deleted because of the existence of FK constraints. PrivateSQL enables privacy to be designated at multiple resolutions. The data owner can flexibly designate which entities in the schema need privacy. The key idea is that one relation is specified to be the primary private relation. Privacy protection extends to additional private relations linked to the primary one via FKs, which are called secondary private relations.

Under this DP policy, two database instances are considered neighbors when one can be obtained from the other by deleting a record x from the primary private relation and cascade deleting other records that refer to x through FKs. One requirement in this approach is that the schema needs to be acyclic. Based on this proposal, researchers began to consider FK constraints when defining neighboring databases [25, 26].

Due to the consideration of FK constraints, the familiarity of cascade deletion on which its neighborhood concept relies, and the general adoption of the model presented in [24], we adopt this model as the starting point for O2 and aim at providing an equivalent model in RDF. The related formal distance definition will be restated in our model in Definition 9.

3   Setting: Formalizing the Concepts

This section introduces an illustrative example based on a Twitter dataset that will be used in the remainder of the paper. It then proposes the formalization of RDB, graphs representing RDF databases, and the mapping from RDB to RDF.

3.1   Illustrative Example

Illustrative Dataset. In this paper, we use as an illustrative example a simple Twitter database, inspired by the Sentiment140 dataset composed of 1.6 million tweets[3]. Its ER diagram is presented in Fig. 1 and an instance of the database is illustrated in Fig. 2.

---

[3] https://www.kaggle.com/kazanova/sentiment140.

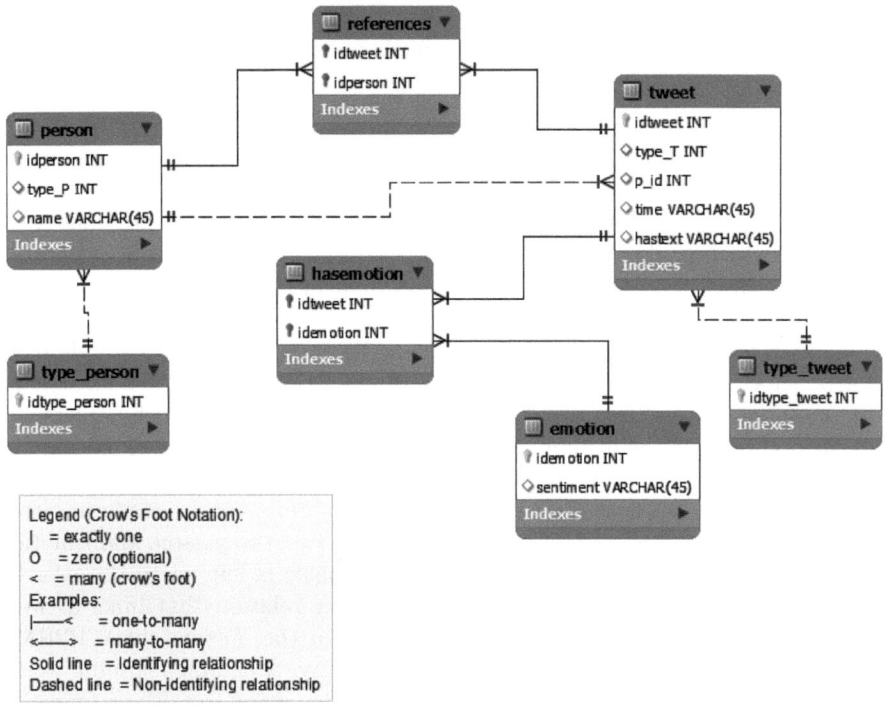

Fig. 1. Entity-Relation Schema of the Sentiment140 dataset

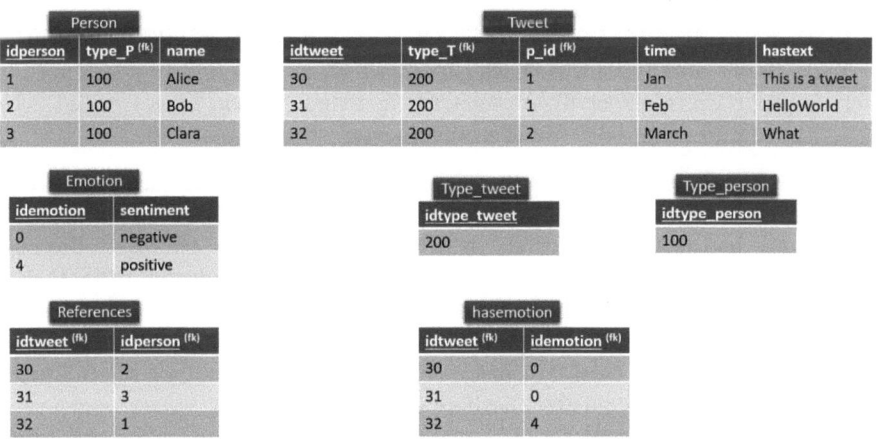

Fig. 2. A database instance of the schema in Fig. 1.

This example presents two many-to-many relationships: 1) between Person and Tweet, captured by the References table with FK referencing the id of a tweet and the id of the persons it references; 2) between Tweet and Emotion, captured by the HasEmotion table. The Tweet table possesses a FK "p_id" from table Person referencing the person that authored the tweet.

Our Mapping Process. As we aim to compare mechanisms and DP models in RDB and RDF, we require a framework in which we can specify encodings, or mappings, from one to the other. We select R2RML-F [27], an R2RML implementation available on Github[4]. The R2RML mapping process is done with R2RML-F whose engine takes as input the RDB, the R2RML mapping file that dictates the direction of relations, and the format of the output file to generate RDF data.

R2RML represents one-to-many and many-to-many relations as simple RDF triplets, which is to say edges in our graphs. Since RDF is an oriented formalism, such mappings require user input to specify in which directions such triplets will end up pointing. We will manually write R2RML mapping files exploring the consequences of those decisions. For instance, in Fig. 3a, the mapping of many-to-many relations, such as *References* and *HasEmotion*, can be made in either direction. We choose that Tweet references go from tweet to person, and Emotion labelling goes from tweet to emotion, but this choice is left up to the user by the W3C recommendations[5]. For the one-to-many relation that links a tweet to its author (modeled by the foreign key *p_id* in the *Tweet* table). R2RML also allows for this relation to be oriented either way. A default mapping would follow W3C recommendations to model it by an edge going from the referencing table (here, *Tweet*) to the referenced table (here, *Person*) as depicted in Fig. 3a. However, for our considerations, we will often consider non-default mappings that do not follow this rule, for instance in Fig. 3b, which depicts a fragment of another encoding of this database, where the edge representing the relation between *Tweet* and *Person* is "backwards", which we will find to sometimes be more appropriate (see Sect. 6). As such, the fact that "person 2 has tweeted tweet 32" is translated as an edge going from r:person_2 to r:tweet_32.

The R2RML-F Engine takes the Twitter relational database and the R2RML mapping file as input and generates the output file in turtle format (.ttl) available online[6]. We use R2RML-F for the R2RML mapping code. However, the code (from 2012), needed some updates. Our 2025 runnable code for R2RML-F is available at: https://github.com/sarataki/mapping/tree/main/code.

Throughout this paper, IRIs are simplified using prefixes. In Listing 1.1 we present all the prefixes used. For instance "rdf:" is a shorthand for the full prefix "http://www.w3. org/1999/02/22-rdf-syntax-ns#".

Listing 1.1. Prefixes

```
PREFIX rdf: <http://www.w3.org/1999/02/22-rdf-syntax-ns#>
PREFIX tweet: <http://foo.example/DB/tweet/>
```

---

[4] https://github.com/chrdebru/r2rml.
[5] https://github.com/sarataki/mapping/tree/main/defaultR2RML/r2rml.ttl.
[6] https://github.com/sarataki/mapping/tree/main/defaultR2RML/output.ttl.

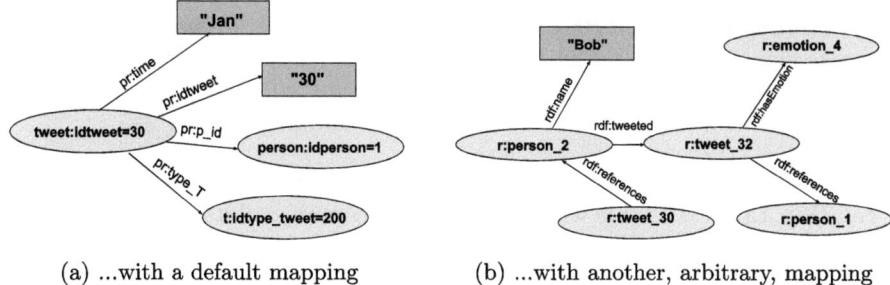

(a) ...with a default mapping   (b) ...with another, arbitrary, mapping

Fig. 3. Extracts of the database mapped from Fig. 2...

PREFIX person: $<$http://foo.example/DB/person/$>$
PREFIX references: $<$http://foo.example/DB/references/$>$
PREFIX reference: $<$http://foo.example/DB/references#$>$
PREFIX pr: $<$http://foo.example/DB/tweet#$>$
PREFIX t: $<$http://foo.example/DB/type_tweet/$>$
PREFIX r: $<$http://example.com/resource/$>$

3.2   Relational Database

We use a conventional notion of schema for relational databases.

Definition 3 (Database Schema).

- A **table schema** T is a set of attribute names.
- A **primary key constraint** PK_T on T is a subset of the attributes of T.
- A **foreign key constraint** ϕ_{l_0,l_1} from T_0 to T_1 is a pair of equal-length lists (l_0, l_1) of attributes of T_0 and T_1 respectively.
- A **database schema** $\mathcal{D}$ is a finite set of tables $\mathcal{T}$ and of constraints $\mathcal{C}$ such that each table T has exactly one primary key constraint PK_T in $\mathcal{C}$, and for all foreign key constraint ϕ_{l_0,l_1} from T_0 to T_1, $PK_{T_0} = l_0$.

We assume tables and attribute sets are disjoint, and that attributes do not have types or domains for this formalism.

Definition 4 (Database). A **database** following a certain database schema $\mathcal{D} = (\mathcal{T}, \mathcal{C})$ is a set D of elements x (called records) such that:

- x belongs to exactly one table $T \in \mathcal{T}$, we note $x \in T$
- For each s an attribute of T, we note $x.s$ the value of attribute s for x. If it is undefined, we say $x.s = $ null. By extension, if l is a n-uplet of attributes, $x.l$ is the n-uplet of their values in x.

The constraints $C \in \mathcal{C}$ are interpreted as such:

- **Primary key** For all PK constraint PK_T, for all $x \in T$, for each $s \in PK_T$, $x.s$ is defined, and if $x \neq x'$ then there exists $s \in PK_T$ such that $x.s \neq x'.s$.
- For all FK constraint ϕ_{l_0, l_1} from T_0 to T_1, for all $x' \in T_1$, there exists a unique element $x \in T_0$ called its antecedent such that $x.l_0 = x'.l_1$.

In this setting, we identify a common type of tables in relational databases. A **relation table** is a table whose primary key contains all its attributes, and is composed of the disjoint union of the domain of two foreign keys. Other tables are called **entities**.

In the classical entity-relational model, many-to-many relations are stored in relation tables, whereas one-to-many relations are directly embedded in entities through foreign keys. In our example database, the relation "tweeted" is a one-to-many, as each tweet has only one author, and is thus stored as a foreign key in the Tweet table. However, the "references" relation is many-to-many, as a tweet can reference several people, and a person can be mentionned in several tweet. The table References stores this relation as pairs of foreign keys from Person and Tweet.

We say that a database schema $\mathcal{D} = (\mathcal{T}, \mathcal{C})$ is **ER-compliant** if for all FK constraints ϕ_{l_0, l_1} from some T_0 to some T_1:

- T_0 is an entity
- either T_1 is a relation and all attributes of l_1 are in its primary key PK_{T_1}, or T_1 is an entity and no attribute of l_1 is in its primary key PK_{T_1}.

3.3   Graph Database and Distances

We present a brief definition of graph databases. It is classical, but distinguishes attributes in a way that will facilitate encodings between RDB and RDF. A RDF dataset is represented as a graph:

Definition 5 (Graph Database). *A graph database is a tuple (A, L, V, E) such that:*

- *A is a potentially infinite set of attribute values*
- *L is a potentially infinite set of edge labels*
- *V is a finite set of vertices*
- *E ⊆ V × L × (V ∪ A) is a set of edges. In an edge v, l, v' we call v the subject, l the predicate, v' the object.*

In this definition, RDF triples are modeled as edges, either between two nodes or from a node to one of its attributes. Attributes here correspond to literals in RDF: they cannot be the subject of a relation and cannot appear isolated. We note that those attributes are not nodes themselves, and will not be counted as such in future distances. L denotes all possible predicates and A denotes the domain of definition of literals that may be object of a predicate.

In the figures, by convention, we represent nodes as yellow ovals and attributes as red rectangle. For example, in Fig. 3b, the node "r:person_2" has an out-edge labeled "rdf:name" whose destination is an attribute "Bob". This represent an RDF triple whose object has URI "r:person_2", with predicate "rdf:name" and object the literal string "Bob". From the remainder of the graph, we see that the individual named Bob is the author of "r:tweet_32" which references "r:person_1", etc.

3.4   Encoding Formalizing a Mapping

The default mapping of the RDB presented in Fig. 2 is illustrated in Fig. 2. To produces another mapping, one may manually write R2RML mappings, which are tailored to their database schema. The choice of an encoding, or mapping, from RDB to graphs modeling a RDF dataset, can be characterized as picking a direction for all those triples. In our formalism, it is defined as follows:

Definition 6 (Orientation). *Let* $\mathcal{D} = (\mathcal{T}, \mathcal{C})$ *be an ER-compliant database schema. An orientation* σ *is a function that associates to each relation table and foreign key between entities a direction.*

- *For all* T *relation table, there exists two FK constraints* $\phi_{l_0,l}$ *from* T_0 *to* T *and* $\phi_{l_1,l'}$ *from* T_1 *to* T *such that* l, l' *are a partition of the attributes of* t. *An orientation for* T *is then* (l, l') *(from* l *to* l') *or* (l', l) *(from* l' *to* l). *We can also note these orientations* (l_0, l_1) *and* (l_1, l_0), *or* (T_0, T_1) *and* (T_0, T_1) *if* $T_0 \neq T_1$.
- *For all FK constraint* ϕ_{l_0,l_1} *from entity* T_0 *to entity* T_1, *an orientation is* (l_0, l_1) *(from* l_0 *to* l_1) *or* (l_1, l_0) *(from* l_1 *to* l_0). *We can also note these orientations* (T_0, T_1) *and* (T_0, T_1) *if* $T_0 \neq T_1$.

Definition 7 (ER encoding). *Let* $\mathcal{D} = (\mathcal{T}, \mathcal{C})$ *be an ER-database schema, and* σ *an orientation. An encoding of relational databases following* $\mathcal{D}$ *into graphs is an injective function* f *from the set of all relational databases following* $\mathcal{D}$ *into the set of graphs, that matches all ER-database* D *following* $\mathcal{D}$ *a graph* $f(D) = (A, L, V, E)$ *such that:*

- **Nodes:** *For each entity* T, *for each* $x \in T$ *in* D, *there is a node* $x \in V$
- **Labels:** *The labels* L *are the union of*
  - *The attributes* s *of all the tables of* $\mathcal{T}$
  - *The relation tables*
  - *The entity-to-entity foreign key constraints*
- **Attributes:** *For each entity* T, *for each attribute* s *such that for no FK constraint* ϕ_{l_0,l_1}, $s \in l_1$, *for each* $x \in T$ *with a defined value for* s $x.s = a$ *in* D, *there exists a value* $a \in A$ *and an edge* $(x, s, a) \in E$
- **Many-to-many relations:** *For each relation* T, *its two FK constraints* $\phi_{l_0,l}$ *from* T_0 *to* T *and* $\phi_{l_1,l'}$ *from* T_1 *to* T *such that* l, l' *are a partition of the attributes of* t, *of orientation* $\sigma(T) = (l, l')$, *for all* $x \in T$, $y \in T_0$ *the antecedent of* $x.l$, *and* $z \in T_1$ *the antecedent of* $x.l'$, *there is an edge* $(y, T, z) \in E$

- ***One-to-many relations:*** *For all FK constraint ϕ_{l_0,l_1} from entity T_0 to entity T_1, for all $x \in T_1$, $y \in T_0$ the antecedent of $x.l_0$, if $\sigma(\phi_{l_0,l_1}) = (l_0,l_1)$ there is an edge $(y, \phi_{l_0,l_1}, x) \in E$, if $\sigma(\phi_{l_1,l_0}) = (l_1,l_0)$ there is an edge $(x, \phi_{l_0,l_1}, y) \in E$*
- *There is no other node, attribute, or edge in $f(D)$*

In the rest of this paper, we will consider distances and how they are preserved through encodings. An ER encoding is an **isometry** w.r.t. a distance d on relational databases and a distance d' on graph databases iff for all D, D' relational databases following $\mathcal{D}$, if $d(D, D')$ is defined then $d'(f(D), f(D'))$ is defined and equal to $d(D, D')$.

For our recurring example, one possible encoding of some entries of Fig. 2 is the graph of Fig. 3b. The R2RML mapping file[7] and the output file are available online[8]. The entities (e.g. person 2, tweet 32, emotion 4) are translated into nodes. However, relations such as References and foreign key relations such as the *p_id* foreign key in the *Tweet* table become labellings in L and are represented as edges, e.g. (r:tweet_32,rdf:references,r:person_1).

4   Distances and Isometries for DP in RDB and RDF

In this section, we formalize two classic distances, namely cascade deletion distance in RDB (Sect. 4.1) and node distance in RDF (Sect. 4.2). Because encodings transform entities into nodes, the cascade distance allows the deletion of several entities at once, but the node distance does not allow the deletion of several nodes at once, no encoding is an isometry between both distances in the general case. We present on the one had a RDF distance that matches cascade deletion (Sect. 4.3), and on the other hand a RDB distance that matches the graphs' node distance (Sect. 4.4).

4.1   Cascade Deletion in RDB

We first present a generalized notion of cascade deletion. Then we show the special case that corresponds to the transitive deletions introduced by Kotsogiannis et al. [24], and we analyze this notion on the RDB example, Twitter.

In general, a cascade deletion is the repercussion of the deletion of an element in a table to all other elements that depended on it in others. The characterization of such dependencies usually revolves around foreign keys, but may vary from a formalization to another. For this reason, we define here the cascade deletion as parameterized by its dependencies.

Definition 8 ($\mathcal{C}'$ Cascade Deletion). *Let D be a database on a schema $\mathcal{D} = (\mathcal{T}, \mathcal{C})$, and $\mathcal{C}' \subseteq \mathcal{C}$ a set of FK. Let x be an element of D. The cascade deletion of x alongside $\mathcal{C}'$ defines a set of **deleted elements** $L_{rm(x)}$ as the smallest set of lines such that:*

---

[7] https://github.com/sarataki/mapping/tree/main/propR2RML/r2rml.ttl.
[8] https://github.com/sarataki/mapping/tree/main/propR2RML/output.ttl.

- $x \in L_{rm(x)}$
- If z is an element in T_1, such that there exists a foreign key $\phi_{l_0,l_1} \in C'$ from T_0 to T_1, and the antecedent of z by ϕ_{l_0,l_1} is $y \in L_{rm(x)}$, then $z \in L_{rm(x)}$.

This is turn defines a set $A_{rm(x)}$ of **deleted attributes**, pairs (y, s) such that:

- $y \in T_1$ is not in $L_{rm(x)}$
- there exists a foreign key constraint $\phi_{l_0,l_1} \notin C'$ from T_0 to T_1
- the antecedent of y by ϕ_{l_0,l_1} is in $L_{rm(x)}$

The result of the cascade deletion of x in D is a database D' of schema $\mathcal{D} = (\mathcal{T}, \mathcal{C})$ whose elements are all the elements of D that are not in $L_{rm(x)}$ where for every $(y, s) \in A_{rm(x)}$, $y.s$ is set to null.

We use this deletion formalism to define a distance:

Definition 9 (C'Cascade Distance). Let D, D' be two databases of same schema, and C' a set of entity to entity FK. We say that D and D' are C' Cascade neighbors if D' is the result of the cascade deletion of an element x in D alongside C', or D is the result of the cascade deletion of an element x in D' alongside C'.
The C' cascade distance is defined over databases of same schema as the length of a shortest path connecting two databases neighbor by neighbor, if it exists.

We note that this definition is "eager" in its deletion, which is to say elements are deleted as soon as one of their relevant antecedents is deleted. There exists another, "cautious" (or lazy) cascade deletion, where elements get deleted only if all their relevant antecedents are deleted. The definition of transitive deletions [24] uses this eager deletion strategy. Our own proposed distance (Sect. 4.4) will circumvent the problem by limiting deletions in a way this distinction no longer matters.

We also note that if C' does not contain all FK of relation tables, it is possible to have "dangling" relation records that potentially break primary key unicity. For simplicity's sake, we will focus on the cases where it does not happen.

4.2 Classical RDF Node Distance

In graphs, records in entities are encoded as nodes. As such, the classical measure most closely related to cascade deletion of a record is the node distance.

Definition 10 (Node Deletion). Let $G = (A, L, V, E)$. We call E_{rmv} the set of edges incident to v: $(v_0, l, v_1) \in E_{rmv}$ iff $v_0 = v$ or $v_1 = v$. The result of the node deletion of v in G is a graph $G' = (A, L, V \backslash \{v\}, E \backslash E_{rmv})$.

Accordingly, the transposition of the node distance in this formalism is:

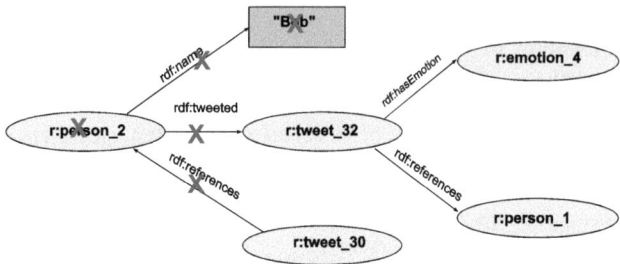

Fig. 4. Node deletion.

Definition 11 (Node Distance). *Let* $G = (A, L, V, E)$ *and* $G = (A, L, V', E')$ *be two graph databases. We say that* G *and* G' *are node-distance neighbors if* G' *is the result of the deletion of a node* $v \in V$ *in* G, *or* G *is the result of the deletion of a node* $v \in V$ *in* G'.

The node distance is defined over graphs of same labels as the length of a shortest path connecting two databases neighbor by neighbor, if it exists.

Figure 4 illustrates the deletion of node "r:person_2" from the graph pictured in Fig. 3b. All the edges incident to it are deleted (those labeled "rdf:name", "rdf:tweeted", and "rdf:references"). While still formally in A, the attribute "Bob" does not appear in the graph anymore since the triple it was the object of has been suppressed. We recall that literals cannot appear while isolated but do not count this as the suppression of a node toward the distance. The resulting graph, containing 4 nodes and two edges, is a node-neighbor of the original graph.

4.3   Cascade Deletion Distance in Graphs

It is immediate that in the general case, encodings cannot hope to be isometric from the RDB cascade distance to the RDF node distance. Indeed, some RDB cascade deletions lead to the deletion of several records in entity tables. In an encoding, this would translate as the deletion of several nodes. However, the RDF node distance only allows the deletion of one node at a time.

We present herein the graph distances and encodings for which this isometry holds. In such cases, comparisons of privacy guaranteeing mechanisms across formalisms would be possible. The definition of a cascade deletion in graphs would follow certain labels as follow:

Definition 12. *Let* $G = (A, L, V, E)$ *be a graph database and* $L' \subseteq L$ *a set of labels. Let* x *be an element of* V. *The cascade deletion of* V *alongside* $\mathcal{L}'$ *defines a set of* **deleted nodes** $V_{rm(x)}$ *as the smallest set of nodes such that:*

- $x \in V_{rm(x)}$
- *If* $l \in L'$, $y, z \in V$ *such that* $(y, l, z) \in E$, *and* $y \in V_{rm(x)}$, *then* $z \in V_{rm(x)}$.

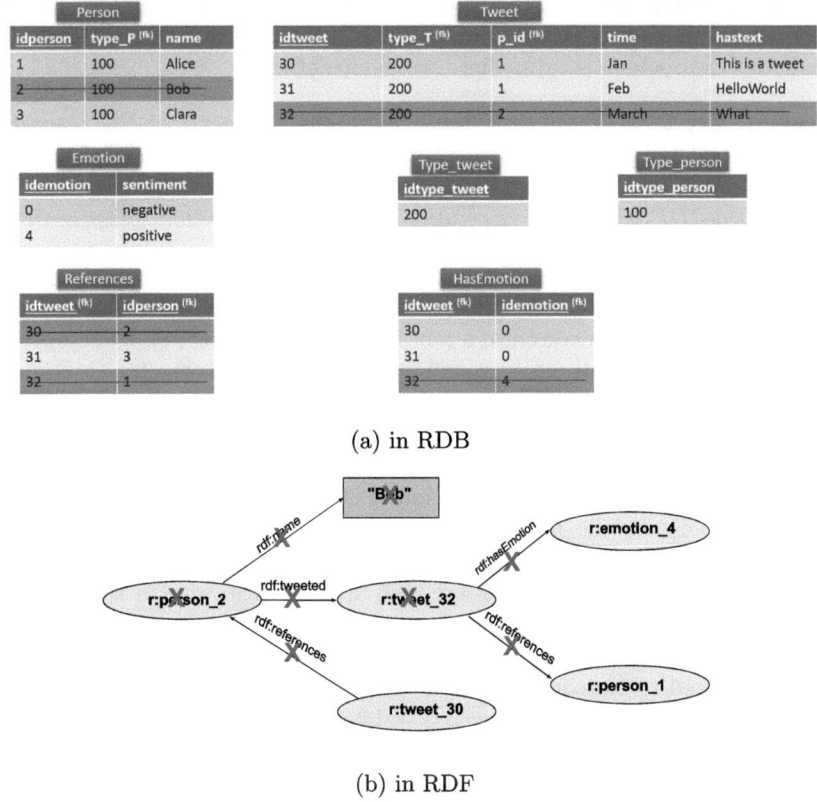

Fig. 5. Transitive cascade deletion with Person as primary entity.

This is turn defines a set $E_{rm(x)}$ of **deleted edges**: for $(y, l, z) \in E$, $(y, l, z) \in E_{rm(x)}$ iff $y \in V_{rm(x)}$ or $z \in V_{rm(x)}$. This also defines a set $A_{rm(x)}$ of **deleted attributes**, $a \in A_{rm(x)}$ if for all y, l such that (y, l, a), $y \in V_{rm(x)}$

The result of the cascade deletion of x along L' in G is $G' = (A \backslash A_{rm(x)}, L, V \backslash V_{rm(x)}, E \backslash E_{rm(x)},)$

Definition 13 (Cascade Deletion Graph Distance). Let $G = (A, L, V, E)$ and $G = (A', L, V', E')$ be two graph databases. We say that G and G' are L'-cascade deletion-distance neighbors if G' is the result of the deletion of a node $v \in V$ along L' in G, or G is the result of the deletion of a node $v \in V$ along L' in G'.

The L'-cascade deletion-distance is defined over graphs of same labels as the length of a shortest path connecting two databases neighbor by neighbor, if it exists.

This cascade deletion works in a way that propagates through some edges of label $l \in L'$. This means that this graph cascade deletion is dependant on the choice of encoding, which is to say on our chosen R2RML mapping. However, for the (natural) choices of encodings and L', this new distance matches RDB cascade deletion, which is immediate by construction.

Lemma 1. *Let $\mathcal{D} = (\mathcal{T}, \mathcal{C})$ be a database schema, $\mathcal{C}'$ a set of FK (containing all relation table FK). We call $L' \subseteq \mathcal{C}'$ the entity to entity FK of $\mathcal{C}'$. Let σ be an orientation such that for all $\phi_{l_0,l_1} \in L'$, $\sigma(\phi_{l_0,l_1}) = (l_0, l_1)$. Then the σ-encoding from RDB to RDF is an isometry from the $\mathcal{C}'$ cascade deletion distance in RDB to the L' cascade deletion distance in RDF.*

For instance, to model cascade deletions in a way that corresponds to [24], we consider that one can compute a join between tables starting from a primary entity T and following all entity to entity foreign key constraints. To study the impact a deletion in the primary table would have on the join, we can delete every line of every table that would no longer occur in it. This corresponds to a transitive deletion alongside those foreign keys.

As an example, the transitive deletion of Bob in the Person table, illustrated in Fig. 5, cascades to the Reference table as tweet 30 can no longer reference him, but also leads to the deletion of his tweet (tweet number 32) which in turns deletes two more lines in the database, one in References, one in HasEmotion.

Limitations. We now discuss the two limitations of cascade deletion as presented here as compared to Kotsogiannis *et al.* [24]. First, for the approach of [24], the choice of a primary table is restricting. While the join approach and transitive deletion as described in [24] necessitates picking a starting point, this has the undesirable side-effect of locking the privacy model towards certain protections and away from others. In RDF, it is possible to use a privacy model protecting any node (DP with node distance) or even nodes from one or several tables exclusively (DP with type-node distance). This is not always possible in databases once a primary table is picked. For instance, in the given database, if we pick Person as the primary table, Fig. 5a shows the only possible way to delete tweet 32. It is then impossible to delete a specific tweet without deleting its author and all its other tweets. In turn, choosing Tweet as a primary table would make it impossible to protect a Person. In a privacy setting, this a restriction that (typed) node DP would not exhibit.

Furthermore, cascade deletion can have a greater or lesser impact based on the chosen starting table. To illustrate this point, in Fig. 6, we show a cascade deletion starting from a tweet. In the corresponding RDF graph, this leads to the deletion of a single node and all its adjacent edges, which is coherent with a node distance of 1. However, in Fig. 5, we show a cascade deletion starting from a person. In the corresponding RDF graph, this leads to the deletion of two nodes and their adjacent edges, which is coherent with a node distance of 2. While it is possible to define an equivalent distance in graphs and propose DP mechanisms accordingly, they would be at risk of having low utility. Indeed, the number of nodes affected by a single deletion being unbounded is a problem in a DP setting, as it aims to guarantee a protection between neighbors. Providing node-DP while maintaining acceptable utility can be challenging, and in the present case neighboring database would differ even more.

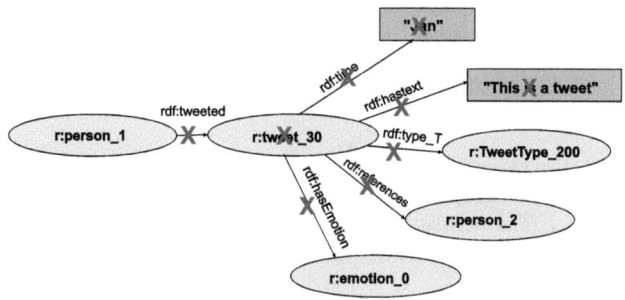

Fig. 6. Transitive cascade deletion of a tweet as shown on a graph.

4.4 A New Meaningful Distance: Restrict Cascade Distance

The previous subsection resolves the mismatch between cascade deletion distance in RDB and node distance in RDF by creating an RDF distance that matches cascade deletion distance. Conversely, we can reverse the translation and search for an RDB distance that matches the node distance. We propose another instance of the cascade deletion: the **restrict cascade deletion**. The key idea is that the deletion of elements from an entity only propagates on the neighboring relations.

Definition 14 (Restrict Cascade Deletion). *Let D be an ER-compliant database on a schema $\mathcal{D} = (\mathcal{T}, \mathcal{C})$, $T \in \mathcal{T}$ a table, and $x \in T$ be an element of D. The restrict cascade deletion of x in D is its $\mathcal{C}'$ cascade deletion, where $\mathcal{C}'$ is the set of FK of relation tables.*

Definition 15 (Restrict Cascade Distance). *Let D, D' be two ER-compliant databases of same schema. We say that D and D' are Restrict Cascade neighbors if D' is the result of the cascade deletion of an element x in D, or D is the result of the cascade deletion of an element x in D'.*
The Restrict cascade distance is defined over ER-compliant databases of same schema as the length of a shortest path connecting two databases neighbor by neighbor, if it exists.

The restrict cascade deletion behaves similarly to the transitive distance except that in some cases it accepts a *null* as FK rather than deleting the concerned line. For example, the restrict cascade deletion is exactly equivalent to the transitive deletion in Fig. 6. Compared to the deletion performed in Fig. 5, restrict cascade deletion is gentler. It would replace the FK p_id with value 2 by a *null*, but preserve the line, and the process would then stop, rather than leading to the suppression of lines in the tables References and HasEmotion.

Notably, this meaningful notion of distance in the relational world, is, no matter the starting table or the data we are trying to protect, always isometric to RDF node distance , as node deletion is exactly∅-cascade deletion.

Theorem 1. *All ER encodings are isometric w.r.t the restrict cascade deletion distance and the node distance*

The proof of this theorem is made by establishing the following lemma:

Lemma 2. *In an ER database, restrict cascade deleting an element of an entity is exactly deleting adjacent relations and erasing adjacent foreign keys.*

Proof (Lemma). Restrict cascade deletion only propagates through foreign keys from an entity to a relation, hence the first part of the lemma: every element of a relation pointing towards the original element are deleted, and it does not propagate further. Erasure concerns foreign keys coming from deleted nodes. However, the only deleted nodes are the original and adjacent relations. Since in an ER database, foreign keys only come from entities, the foreign keys coming from deleted nodes all come from the original. □

The proof of the theorem is a direct consequent:

Proof (Theorem). Let D be a database, x one of it's elements, D' the database resulting from the cascade deletion of x in D, and f an ER encoding. D and D' are neighbors in the restrict cascade deletion distance, from Lemma 2. We will prove $f(D)$ and $f(D')$ are neighbors in the node deletion distance. The only difference between D and D' is:

- The deletion of x: this translates as the disappearance of the node x from V and the deletion of all its information, which translates as a deletion of all edges outgoing from x between $f(D)$ and $f(D')$.
- The deletion of every adjacent relation element: this translates as the disappearance of some edges between x and other elements of V that corresponds to the encoding of relations between $f(D)$ and $f(D')$.
- The erasure of every foreign key coming from x: this translates as the disappearance of all remaining edges between x and elements of V that corresponds to the encoding of entity-to-entity foreign keys between $f(D)$ and $f(D')$.

As a summary, between $f(D)$ and $f(D')$, we have removed x from V, and all edges incident to x from E. We note that this is the exact definition of the node deletion of x in $f(D)$, and conclude that $f(D)$ and $f(D')$ are neighbor in the node deletion distance.

Note that this argument goes both ways: any node deletion in $f(D)$ would result in a new graph which is the encoding of another relational database D'', which is identical to D save for one restrict cascade deletion.

Since both the cascade deletion distance and the node deletion distance are defined as the shortest distance from neighbor to neighbor between two points, this preservation of neighborhood is sufficient to prove that f is an isometry. □

5 QL-Outedge Privacy over Relational Databases

In Sect. 4, we proposed a meaningful adaptation of a classical privacy model in relational databases, that when transformed to an RDF database, correspond to typed-node privacy. However, as discussed in Sect. 2, other privacy models on graphs are better adapted to RDF graphs, such as QL-Outedge privacy [8]. Thus in this section, we propose to study how we can translate the QL-Outedge privacy model to the relational database setting. The final goal is to propose a neighborhood definition for a relational database which would be the equivalent of the QL-Outedge neighborhood defined over RDF graphs.

Our approach is to map the relational database to RDF, apply QL-Outedge privacy in RDF, then return to relational world to define the corresponding neighborhood. When mapping a relational database to RDF, there is the edge direction to be tailored in order to provide the appropriate privacy guarantee under QL-Outedge privacy. Thereupon, we propose a model where one can choose the direction of edges according to what to protect.

Desired Privacy and Motivating Queries. As a way to illustrate the importance of varied (and matching) distances between RDB and RDF formalisms, we consider a case where the desired protection of privacy is on the entire set of tweets of a single person. As such, any sufficiently private mechanism should be noisy enough for an attacker not to be able to significantly distinguish between the inclusion or exclusion of the set of tweets authored by a person from the mechanism's result. As a measuring stick, we study DP mechanisms that would answer the following queries while adding a noise calibrated by the global sensitivity of those queries under a particular distance.

- **Motivating Example 1:** Consider query Q1: Find the maximum number of tweets tweeted by a single person (maximum out-degree of *tweeted* outedges).
- **Motivating Example 2:** Consider query Q2: Count how many users "Alice" has referenced.

5.1 QL-Outedge Distance

We express the definition of QL-Outedge distance in our formalism.

Definition 16 (QL-Outedge pruning). *Let $G = (A, L, V, E)$. We call O_{QLv} the set of edges coming from v of label in QL: $(v, l, v') \in O_{QLv}$ iff $l \in QL$. The result of the QL pruning of v in G is a graph $G' = (A, L, V, E \backslash O_{QLv})$.*

Accordingly, the transposition of the node distance in this formalism is:

Definition 17 (QL-Outedge Distance). *Let $G = (A, L, V, E)$ and $G = (A, L, V, E')$ be two graph databases. We say that G and G' are QL-Outedge neighbors if G' is the result of the QL pruning of a node $v \in V$ in G, or G is the result of the QL pruning of a node $v \in V$ in G'.*

The QL-Outedge distance is defined over graphs of same labels as the length of a shortest path connecting two databases neighbor by neighbor, if it exists.

We note that only graphs of identical sets of nodes have a definable distance in QL-Outedge.

5.2 RDB Keywise Deletion

Similarly to the previous section, we aim to find an RDB distance that matches the RDF QL-outedge distance. To do so, we will define keywise deletion as a direct analog of QL-outedge pruning.

Definition 18 (Relational Keywise Pruning). *Let D be a ER-compliant database on a schema $\mathcal{D} = (\mathcal{T}, \mathcal{C})$, and $\gamma \subseteq \sigma$ a restriction of an orientation σ to part of its domain. Let x be an element of T an entity table of $\mathcal{T}$. The relational keywise pruning of x alongside $\mathcal{C}', \sigma$ defines a set of **deleted elements** $L_{KW(x)}$ as the set of lines z such that:*

- *z is in a relation table T' of FK $\phi_{l,l_0'}$ from T to T' and $\phi_{l'',l_1'}$ from some T'' to T',*
- *$\gamma(T') = (l_0', l_1')$,*
- *x is the ϕ-antecedent of z for $\phi_{l,l_0'}$.*

*The set $A_{KW(x)}$ of **deleted attributes** is the set of pairs (y, s) such that:*

- *$y \in T'$ where T' is an entity table*
- *there exists a FK constraint $\phi_{l,l'}$ from T to T'*
- *$\gamma(\phi_{l,l'}) = (l, l')$*
- *the antecedent of y by $\phi_{l,l'}$ is x*
- *$s \in l'$*

or (x, s) such that:

- *there exists a FK constraint $\phi_{l',l}$ from T' to T*
- *$\gamma(\phi_{l',l}) = (l, l')$*
- *$s \in l$*

The result of the relational keywise deletion of x alongside γ is a database D' of schema $\mathcal{D} = (\mathcal{T}, \mathcal{C})$ whose elements are all the elements of D that are not in $L_{KW(x)}$ where for every $(y, s) \in A_{QL(x)}$, $y.s$ is set to null.

γ represents both a choice of relations, many-to-many or one-to-many (the restricted domain) and the only direction from which such relation must be pruned. We note that in one-to-many relationships, the resulting deletion of attributes can happen in a record away from x (if the orientation goes from the antecedent outward) or in x itself (if the orientation goes from the postcedent to the antecedent). We also note that both orientations cannot be picked for a single relation, mirroring a known limitation of QL-outedge.

Definition 19 (Relational Keywise Distance). *Let D, D' be two database of same schema, and γ a restriction of an orientation. We say that D and D' are relational keywise neighbors if D' is the result of the relational keywise deletion of an element x in D alongside γ, or D is the result of the relational keywise deletion of an element x in D' alongside γ.*

The relational keywise deletion distance is defined over databases of same schema as the length of a shortest path connecting two databases neighbor by neighbor, if it exists.

We note that only databases of identical entities have a definable distance in Keywise Distance, up to attribute deletion, as entity lines are never part of $L_{KW(x)}$.

5.3   Isometric Encoding RDB to RDF for QL-Outedge

We now characterize under which conditions the keywise distance and QL-outedge distance are linked by an isometric encoding. Since all entities/nodes remain untouched, and edge orientation is at the heart of QL-Outedge, this property will focus on finding the appropriate orientation.

Lemma 3 (QL-outedge isometry). *Let $\mathcal{D} = (\mathcal{T}, \mathcal{C})$ be a database schema, γ an orientation restriction, σ an orientation, QL a set of labels.*

The σ-encoding of databases of schema $\mathcal{D}$ is an isometry between γ-keywise distance and QL-outedge distance if γ is the restriction of σ on QL.

Proof. Let D be a database, and $f(D)$ its σ-encoding. Let x be an element of an entity T of D. It is a node in $f(D)$. This node's outgoing edges come from three sources:

- **Many-to-many relations:** the lines of D encoded as edges to or from x are lines of relations table that have x for antecedent in one of their FK. Hence the outgoing edges of x come from the set of lines z such that:
  - z is in a relation table T' of FK ϕ_{l,l'_0} from T to T' and ϕ_{l'',l'_1} from some T'' to T',
  - $\sigma(T') = (l'_0, l'_1)$,
  - x is the ϕ-antecedent of z for ϕ_{l,l'_0}.
- **One-to-many relations:** the entity to entity FK of D encoded as edges to or from x are those where x is the antecedent or postcedent. Depending on the role x takes and the chosen orientation, those will be incoming or outgoing. The outgoing ones specifically are the set of pairs (y, l') such that:
  - $y \in T'$ where T' is an entity table
  - there exists a FK constraint $\phi_{l,l'}$ from T to T'
  - $\sigma(\phi_{l,l'}) = (l, l')$
  - the antecedent of y by $\phi_{l,l'}$ is x

  or (x, l) such that:
  - there exists a FK constraint $\phi_{l',l}$ from T' to T
  - $\sigma(\phi_{l',l}) = (l, l')$

- $s \in l$

If we were to prune x in $f(D)$ along QL, we would restrict all of those edges to those that fulfill those condition and are from a relation or FK in QL. That is to say, it will be the same specification but every time we check for the orientation σ of a relation or FK, it should also be in QL. In other words, it is the same as checking this equality on the restriction of σ on QL, which is γ. However, replacing σ by γ in the specification above gives us exactly the definitions of $L_{KW(x)}$ and $A_{KW(x)}$. This means that γ-pruning x in D or QL-pruning x in $f(D)$ is identical. This means that finding D' a γ-keywise neighbor of D or finding f(D') a QL-outedge-neighbor of $f(D)$ is identical. This means that the σ-encoding f is an isometry between those two distances.

6   Illustrating Encodings Impact on QL-Outedge Privacy

To illustrate the concept introduced in the previous section and the impact of mapping methods, we refer to our running example dataset and introduce two simple example queries:

Q1. Find the maximum number of tweets tweeted by a single person (maximum out-degree of *tweeted* outedges).
Q2. Count how many users "Alice" has referenced.

We discuss hereafter how protection and global sensitivity differ as different R2RML mappings are proposed. As such, we will present the privacy protection offered by the default mapping. We show that this may lead to weak protection and, given a stronger target, show how to design an appropriate, manually-written, R2RML mapping tailored to the use-case to achieve desired privacy protection. These two different protections are illustrated through Q1 (which they directly relate to) and Q2 (on which they have no impact).

6.1   Default R2RML, Protecting Authorship of a Single Tweet

In this first case, we use a R2RML mapping according to W3C R2RML recommendation[9]. This standard lets the choice of orientation open for many-to-many relationships in our example, References and HasEmotion. We translate both relations as outedges of Tweet. However, for one-to-many relations (between tweets and their author), the standard imposes that we choose a direction from the referencing table (here, the one-side Tweet) to the referenced table (here, the many-side Person). This means that, as is shown in Fig. 3a, the edges of this relation count as outgoing for tweets rather than persons.

The implicit protection provided is that on a QL-outedge distance, a DP mechanism would protect the information of any one tweet. This impacts the neighborhood one can build with QL-outedge, and the resulting global sensitivity of certain queries is affected by this choice:

---

[9] https://github.com/sarataki/mapping/tree/main/defaultR2RML/r2rml.ttl.

Q1. The global sensitivity of this query under QL-Outedge privacy over the RDF graph obtained from default R2RML mapping is 1, assuming that `pr:p_id` $\in$ QL, which represents the *tweetedBy* relation. The neighboring graphs differ by the QL-outedges of an arbitrary node. The *Tweet* node has exactly one *tweetedBy* outedge (along with other outedges of other labels). So, one possible neighboring graph differs by the outedges of node *Tweet*. The model related to this encoding protects the *tweetedBy* outedge, *i.e.,* the author of one tweet.

Q2. The global sensitivity of this query under QL-Outedge privacy over the RDF graph obtained from this default R2RML mapping is infinite, assuming both *tweetedby* and *references* are in QL. The node *Person* has one *name* outedge with Literal value "Alice" (along with other outedges). Any neighboring graphs differ by the QL-outedges of an arbitrary node, possibly a *Tweet*. The *Tweet* node has exactly one *tweetedBy* outedge plus some *references* outedges. Since the number of references outedges could be unbounded, so is this query's global sensitivity.

Discussion: As shown above, the default R2RML mapping restricts our choice on privacy protection and, for example, imposes to protect only the author of one tweet, as exemplified by the sensitivity of Q1 being 1. However, one may want to protect *all* the tweets of a person, of an author, rather than the author of one tweet. Notably, no choice of QL under default mapping would give us the desired graph neighborhood. Under QL-outedge DP, protecting all the tweets of a person requires the edges representing authorship to be outedges rather than inedges of nodes Person. This can be achieved by writing a customized R2RML mapping. We discuss the definition of the related mapping hereafter, as well as the impact of the example queries.

6.2   Custom Mapping, Protecting Someone's Tweets

Our objective is to find a distance for which DP guarantees protection on all the tweets of a person, which is to say a distance where such deletion is possible between a graph and one of its neighbors. As noted above, this is impossible under default mapping. However, by writing another R2RML file where we define our own mappings, we can better control the privacy protection as we reestablish appropriate distance, and consequently, appropriate noise requirement for DP.

Design of a Mapping Satisfying the Target Protection. For our example dataset, protecting all the tweets authored by an individual corresponds in RDB to a keywise distance on the foreign key p_id that points from *Tweets* to *Person*. To delete all pointers to one person in one go, we must choose the orientation that is contrary to the W3C recommendations, that is to say from the referenced table to the referencing table. To find an isometry of this to a QL-outedge distance, Lemma 3 tells us that the encoding we pick must also choose the direction from *Person* to *Tweet* to encode this foreign key. This encoding corresponds to a different R2RML mapping[10], which outputs graphs with properly oriented edges (see Fig. 7). In this new encoding and with the new QL-outedge distance it permits, we reconsider query Q1 and Q2. We study again the global sensitivity with our proposed R2RML mapping.

Q1. The sensitivity of Q1 in this new distance is the maximal number of edges *tweeted* that can separate a graph from its neighbors. This maximal number does not exist, as an arbitrarily great number of edges can be deleted from a single *Person* node at once. The global sensitivity of this query is therefore infinite. This means that we now correctly assess that no amount of noise can obfuscate the presence or absence of a single person's tweet output to the satisfaction of DP criteria.

Note that, if the number of tweets outedges of node person is bounded, then the global sensitivity of Q1 is bounded. This can also be achieved by using a projection method (e.g. [8]).

Q2. Two possible neighbouring graphs differ by the QL-outedges of some nodes. *Tweet* (resp. *Person*) nodes may have an arbitrary number of references (resp. tweeted) outedges. Again, assuming *tweeted* $\in$ QL or references $\in$ QL, we have global sensitivity of Q2 equal to infinite. Here, the change in mapping (and hence privacy protection) does not impact Q2, since the number of people referenced in a single tweet is unbounded, and protecting one or several tweet still lead to the protection of an arbitrarily high number of referenced individuals.

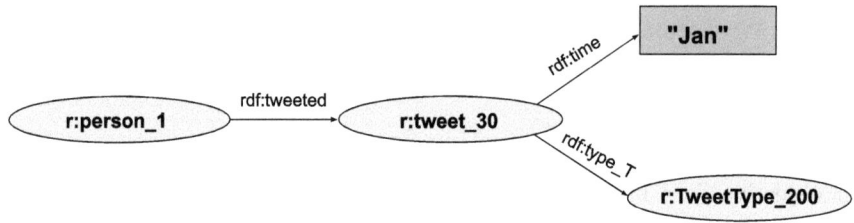

Fig. 7. Proposed R2RML: translation of one-to-many relationship.

Discussion. Hence, the mappings influence privacy protection and, in turn, may impact queries' sensitivity. We note that this choice of orientation is not objectively better, but merely adapted to the protection we use as an example. Every such decision represents some form of tradeoff on what privacy protection can or cannot be expressed. For instance, if we represent authorship information from *Person* to *Tweet*, then privacy protection about a single tweet becomes incongruous: even a complete deletion of all the information outgoing from a *Tweet* node would leave its author edge untouched, as it is now incoming. If translated into RDB, this gives an incomplete line pruning, as shown in Fig. 8.

---

[10] https://github.com/sarataki/mapping/tree/main/propR2RML/r2rml.ttl.

Person			Tweet				
idperson	type_P	name	idtweet	type_T	p_id	time	hastext
1	100	Alice	30	200	1	Jan	This is a tweet
2	100	Bob	31	200	1	Feb	HelloWorld
3	100	Clara	32	200	2	March	What

Fig. 8. Tweet (incomplete) pruning under nonstandard edge orientation.

7 Conclusion

In this paper, we analyze the transposition to RDF through mapping of a popular DP model for multi-tables relational databases with FK constraints related to transitive deletion [24]. We show it has an equivalent in RDF, an extension of node DP, under a choice of R2RML mapping made to correspond to a choice of primary relation. To ease the construction of RDF DP mechanisms while remaining explainable in the relational world, we tweak the original privacy model in a meaningful way so that its translation is always equivalent to classical node DP. Thus, we proposed the restrict deletion for relational databases, which captures privacy policies and FK constraints.

Furthermore, we formalize neighborhoods in relational databases to match the concept of QL-Outedge privacy, which was previously defined over RDF graphs. We believe this approach is particularly suitable to the context of RDF, offering meaningful semantics for the neighborhoods. When mapping relational databases to RDF, we propose a model where one can choose what they want to protect. This is once again done by deciding the edge direction in the R2RML mapping. The decision on which direction to choose will affect what edges are incoming or outgoing, and, therefore, the QL-Outedge neighborhood definition in the relational database. Finally, we proposed an implementation based on R2RML to illustrate our approach.

For future work, we plan to strengthen and implement relational-to-graph and graph-to-relational database mapping methods, by matching known and useful distances of RDB or RDF as well as neighborhood definitions which would make more sense in this context into corresponding notions in the other formalism. Furthermore, another interesting research direction is establishing a benchmark to compare the efficiency of different privacy methods through mapping. This would lead to a wider choice of comparable options for information stored as RDB or RDF while preserving important privacy guaranteeing properties.

Acknowledgments. This work is supported by the French National Research Agency (ANR) under grants iPoP (ANR-22-PECY-0002), CyberINSA (ANR-23-CMAS-0019), and DiffPriPos (ANR-23-CE23-0032).

Disclosure of Interests. The authors have no competing interests to declare that are relevant to the content of this article.

References

1. Field, L., et al.: Realising the full potential of research data: common challenges in data management, sharing and integration across scientific disciplines. Technical report, ZENODO (2013)
2. Michel, F., Montagnat, J., Faron-Zucker, C.: A survey of RDB to RDF translation approaches and tools. Technical report, I3S (2013)
3. Hazber, M.A.G., Li, R., Li, B., Zhao, Y., Alalayah, K.M.A.: A survey: transformation for integrating relational database with semantic web. In: Proceedings of the 2019 3rd International Conference on Management Engineering, Software Engineering and Service Sciences, pp. 66–73 (2019)
4. Boneva, I., Staworko, S., Lozano, J.: Sherml: mapping relational data to RDF (2019)
5. Dwork, C., Roth, A., et al.: The algorithmic foundations of differential privacy. Found. Trends® Theor. Comput. Sci. **9**(3–4), 211–407 (2014)
6. Near, J.P., He, X., et al.: Differential privacy for databases. Found. Trends® Databases **11**(2), 109–225 (2021)
7. Reuben, J.: Towards a differential privacy theory for edge-labeled directed graphs. In: SICHERHEIT 2018 (2018)
8. Taki, S., Eichler, C., Nguyen, B.: It's too noisy in here: using projection to improve differential privacy on RDF graphs. In: Chiusano, S., et al. (eds.) ADBIS 2022. CCIS, vol. 1652, pp. 212–221. Springer, Cham (2022). https://doi.org/10.1007/978-3-031-15743-1_20
9. Taki, S.: Linked Data Sanitization with Differential Privacy. Theses, INSA Centre Val de Loire (2023)
10. Taki, S., Boiret, A., Eichler, C., Nguyen, B.: Cohesive database neighborhoods for differential privacy: mapping relational databases to RDF. In: Barhamgi, M., Wang, H., Wang, X. (eds.) WISE 2024. LNCS, vol. 15440, pp. 231–242. Springer, Singapore (2024). https://doi.org/10.1007/978-981-96-0576-7_18
11. Arenas, M., Bertails, A., Prud'hommeaux, E., Sequeda, J., et al.: A direct mapping of relational data to RDF. W3C Recommend. **27**, 1–11 (2012)
12. Souripriya, D., Seema, S., Richard, C.: R2RML: RDB to RDF mapping language. W3C Recommend. **27** (2012)
13. de Medeiros, L.F., Priyatna, F., Corcho, O.: MIRROR: automatic R2RML mapping generation from relational databases. In: Cimiano, P., Frasincar, F., Houben, G.-J., Schwabe, D. (eds.) ICWE 2015. LNCS, vol. 9114, pp. 326–343. Springer, Cham (2015). https://doi.org/10.1007/978-3-319-19890-3_21
14. Berners-Lee, T.: Relational databases on the semantic web (1998). https://wwww3.org/DesignIssues/RDBRDF.html. Accessed Jan 2015
15. Michel, F., Montagnat, J., Zucker, C.F.: A survey of RDB to RDF translation approaches and tools (2013)
16. Auer, S., Feigenbaum, L., Miranker, D., Fogarolli, A., Sequeda, J.: Use cases and requirements for mapping relational databases to RDF. W3c Working Draft (2010)
17. Dimou, A., et al.: Mapping hierarchical sources into RDF using the RML mapping language. In: 2014 IEEE International Conference on Semantic Computing, pp. 151–158. IEEE (2014)
18. Dimou, A., Sande, M.V., Colpaert, P., Verborgh, R., Mannens, E., Van de Walle, R.: RML: a generic language for integrated RDF mappings of heterogeneous data. Ldow **1184** (2014)

19. Michel, F., Djimenou, L., Zucker, C.F., Montagnat, J.: Translation of relational and non-relational databases into RDF with XR2RML. In: 11th International Conference on Web Information Systems and Technologies (WEBIST'15), pp. 443–454 (2015)
20. Dwork, C.: Differential privacy. In: Proceedings of the Automata, Languages and Programming, 33rd International Colloquium, ICALP (2006)
21. Hay, M., Li, C., Miklau, G., Jensen, D.: Accurate estimation of the degree distribution of private networks. In: 2009 Ninth IEEE International Conference on Data Mining, pp. 169–178. IEEE (2009)
22. McSherry, F.D.: Privacy integrated queries: an extensible platform for privacy-preserving data analysis. In: Proceedings of the 2009 ACM SIGMOD International Conference on Management of data, pp. 19–30 (2009)
23. Johnson, N., Near, J.P., Song, D.: Towards practical differential privacy for SQL queries. Proc. VLDB Endow. **11**(5), 526–539 (2018)
24. Kotsogiannis, I., et al.: PrivateSQL: a differentially private SQL query engine. Proc. VLDB Endow. **12**(11), 1371–1384 (2019)
25. Tao, Y., He, X., Machanavajjhala, A., Roy, S.: Computing local sensitivities of counting queries with joins. In: Proceedings of the 2020 ACM SIGMOD International Conference on Management of Data, pp. 479–494 (2020)
26. Dong, W., Fang, J., Yi, K., Tao, Y., Machanavajjhala, A.: R2T: instance-optimal truncation for differentially private query evaluation with foreign keys. In: Proceedings of the International Conference on Management of Data (2022)
27. Debruyne, C., O'Sullivan, D.: R2RML-F: towards sharing and executing domain logic in r2rml mappings. In: LDOW@ WWW, vol. 1593 (2016)

Author Index

B
Boiret, Adrien   94

C
Colin, Jonathan   1

E
Eichler, Cédric   94

G
Gançarski, Stéphane   32
Gillet, Annabelle   61

J
Jarrad, Sara   32

L
Leclercq, Éric   61

M
Maniu, Silviu   1

N
Naacke, Hubert   32
Nguyen, Benjamin   94

T
Taki, Sara   94

A. Hameurlain et al. (Eds.): *Transactions on Large-Scale Data- and Knowledge-Centered Systems LIX*, LNCS 16240, p. 123, 2026.
https://doi.org/10.1007/978-3-662-72449-1

MIX
Papier aus verantwortungsvollen Quellen
Paper from responsible sources
FSC® C105338

If you have any concerns about our products,
you can contact us on
ProductSafety@springernature.com

In case Publisher is established outside the EU,
the EU authorized representative is:
Springer Nature Customer Service Center GmbH
Europaplatz 3, 69115 Heidelberg, Germany

Printed by Libri Plureos GmbH
in Hamburg, Germany